PLUMBING
& CENTRAL HEATING

MIKE LAWRENCE

The Crowood Press

First published in 1991 by
The Crowood Press Ltd
Gipsy Lane, Swindon
Wiltshire SN2 6DQ

British Library Cataloguing in Publication Data

Lawrence, Mike 1947–
    Plumbing and central heating.
    1. Plumbing – Amateurs' manuals   2.   Dwellings. Heating and ventilation
    I. Title
    696.1

ISBN 1   85223 515 2

Acknowledgements
Line-drawings by Andrew Green.

The author would like to thank the following companies and organizations for
providing the photographs listed below:

Armitage Shanks (pages 39, 43 and 47);
Cristal Tiles (pages 41, 45 and 57);
Marshall Cavendish (pages 7, 35, 37, 49, 51, 53, 61, 63, 65, 67, 85 and 87);
Robert Harding Picture Library (pages 12 to 18);
Triton (page 55).

Typeset by Acūté, Stroud, Gloucestershire.
Printed and bound by Times Publishing Group, Singapore.

Contents

Introduction

Plumbing, like electrical work, is an area where many do-it-yourselfers prefer not to tread. That would be understandable if it were still the traditional and somewhat mystical craft our grandfathers practised, but modern materials and new techniques have made it accessible to the layman. Furthermore, materials and equipment have never been so readily available. There is no longer any need to queue up in the plumbers' merchant and suffer the sneers of the professional while you try to explain what you want; it is all on the shelf of your local DIY superstore. So there is no reason why anyone should not tackle all sorts of plumbing jobs around the house, from simple repairs to major home improvements.

What holds many people back from doing so is fear: fear of making a mess. It is certainly true that plumbing mishaps rank high in do-it-yourself folklore, but it has never been so easy to repair and modify a typical domestic plumbing system. This is thanks to the ingenuity of the equipment manufacturers and the wider availability of accessible information on how to tackle various jobs.

The secret of success – and of conquering your fears about plumbing work – is simple: master the techniques first of all, and then have confidence in your workmanship and in your ability to succeed. This book will help you with the first and encourage you with the second.

THE BASICS

Plumbing work is basically quite straight-forward, but it can be somewhat repetitive; whatever the project, you always wind up doing the same relatively simple operations over and over again. However, that does not make the work any less rewarding, because the end product can have all sorts of benefits. For example, new fittings can improve your bathroom or kitchen dramatically, while new taps can banish drips and plumbing in your washing machine will put an end to all that fiddling with hoses on wash-days.

To become a good do-it-yourself plumber, you need to understand how your plumbing system works, familiarize yourself with the raw materials, master the basic techniques and decide what jobs you can (or want to) undertake. This chapter takes you through all these steps one by one.

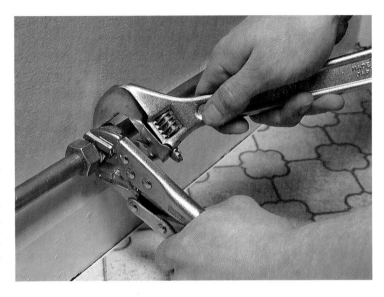

How Water Reaches your Home

Almost every home in the country has the benefit of a mains water supply. The water is distributed through underground mains, and the supply to individual properties is tapped into it beneath the road outside. The supply pipe runs to a control valve, known as the water authority stopcock or stoptap, which is usually located beneath a small metal plate set in the pavement in front of the property.

Under the cover is a shaft, with the stoptap typically about 60cm (24in) below ground level to prevent it freezing in cold

weather. Older buildings may have this cover and shaft located actually on the property, sometimes in the front garden or in a path or drive. The tap generally needs a special key if it is to be operated, and it is well worth obtaining one (from plumbers' or builders' merchants) so you can turn off the supply at this point if necessary.

From this stoptap, the pipe runs on underground to the point where it enters the house, usually through a sleeve to protect it from any settlement of the house wall. It then rises through the floor and terminates in another stoptap, known as the rising main stoptap, which is used to control the flow of all water into the house.

Fig 1 (*above*) All plumbing jobs require mastery of only a few simple techniques.

Fig 2 (*below*) Water reaches the house via an underground supply pipe from the mains in the road.

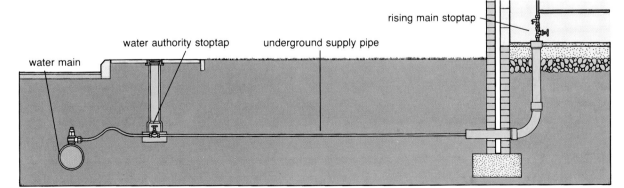

rising main stoptap

water authority stoptap   underground supply pipe

water main

Plumbing, Heating and Waste Systems

Note that the supply pipe is your responsibility from the property boundary onwards. A draincock is usually fitted immediately after the rising main stoptap to allow the supply pipework to be emptied of water for maintenance purposes.

The Cold Water Supply

From the rising main stoptap, the main cold supply pipe rises up through the house (which is why this stoptap is so named). A branch pipe is always connected to the rising main to supply the kitchen cold tap with a pure supply of water for drinking and cooking purposes. There may also be branches off this pipe serving a washing machine or dishwasher.

Indirect systems In the majority of homes the rising main then continues upwards to supply a cold water storage tank in the loft or in an upstairs cupboard. The flow of water into this tank is controlled by a ball valve. As water is drawn off, the valve's float arm drops which opens

up the valve to admit more water. The valve closes again when the tank is filled to the correct level.

This tank then supplies cold water to the rest of the house. Normally there will be two feed pipes running from near the base of the tank; one will supply all the cold taps (except the kitchen) and also WC cisterns. The other will supply cold water to the hot water cylinder. Both pipes should be fitted with an on/off control called a gate valve, which allows the feed to be isolated if necessary.

If the house has a conventional 'wet' central heating system with water-filled radiators, there will be a second, smaller tank in the loft which is supplied by a pipe which branches off the rising main and again which is fitted with a ball valve. This is the heating system's feed-and-expansion tank. Its purpose is to accommodate the expansion in the system's water content as it heats up, and also to replace any losses from the system should they occur. There should be a stoptap on its

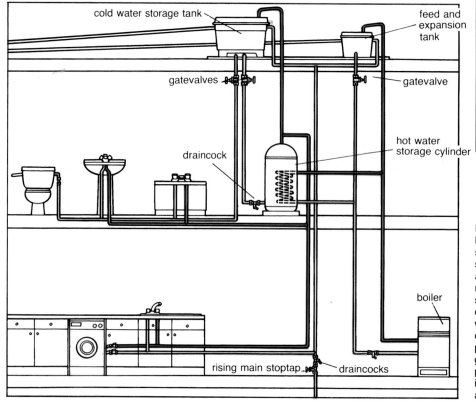

Fig 3 (*left*) A typical household plumbing system, with direct cold supplies to the kitchen tap, automatic washing machine, cold water storage in the loft supplying other cold taps, and indirect water heating via a hot cylinder and boiler. Note the positions of stoptaps and gate valves to allow various parts of the system to be isolated for maintenance or alteration.

branch supply pipe to enable you to turn off its water supply if necessary.

Direct systems Some homes have a direct cold water supply, with branches to taps and WC cisterns taken directly from the rising main. Hot water may be supplied by a multi-point water heater, by a conventional hot cylinder containing an immersion heater, or by a gas or electric storage water heater. Alternatively, there may be a full-scale central heating system which also supplies hot water via a hot cylinder – see below.

Direct plumbing systems may be easier and cheaper to install, but most water authorities prefer indirect systems (and nowadays may not allow direct ones to be installed). One of the main reasons for this is that indirect systems make it very difficult for the mains supply to be contaminated by back-siphonage of dirty water if there is a drop in mains water pressure. This subject is very much to the fore in the current water supply by-laws, and which will be mentioned at intervals throughout this book. Indirect systems are also convenient for the householder, because they guarantee a supply of stored water in the event of an interruption in the mains supply, and they are also quieter in operation than mains-fed systems.

The Hot Water System

As with cold water systems, there are two basic types of hot water system. In one, the water is heated and stored ready for use, while in the other it is heated as it is required.

Storage systems Cold water is supplied (usually from the loft storage tank, but see also page 92) to a copper cylinder, where it is heated and is then piped on to feed the hot taps around the house. The heating may be by means of an electric immersion heater set in the top of the cylinder, or else by a separate boiler. On older *direct systems*, water flows from the cylinder to the boiler and back again, usually relying on gravity for its circulation (warm water is lighter than cold water and so rises and drives the water round the system).

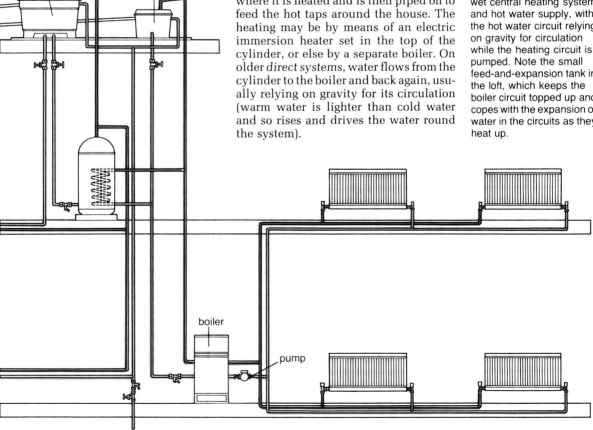

cold water storage tank

feed and expansion tank

boiler

pump

Fig 4 (*below, left*) A typical wet central heating system and hot water supply, with the hot water circuit relying on gravity for circulation while the heating circuit is pumped. Note the small feed-and-expansion tank in the loft, which keeps the boiler circuit topped up and copes with the expansion of water in the circuits as they heat up.

Plumbing, Heating and Waste Systems

The cold supply pipe and the feed to the boiler are connected near the base of the cylinder, and the boiler return pipe is connected near the top. The hot water take-off is from the domed top of the cylinder, and a vent or warning pipe runs up to the loft where it discharges over the cold water tank in the event of the system overheating. This system is not used nowadays because the constant intake of fresh water causes serious scale build-up within the boiler and cylinder.

On *indirect systems*, there is a copper coil inside the cylinder through which circulates water heated by the boiler. The water in the cylinder is heated indirectly by the hot water in the coil; the two never mix so scale build-up in the boiler and coil is minimal. It still occurs on the outside of the coil, but this is less of a problem except in areas with extremely hard water.

Such a system has an additional cold tank in the loft – the feed-and-expansion tank mentioned earlier. Where the boiler also supplies the central heating system, the circuit to the cylinder may rely on gravity for circulation, but on newer systems it is usually pumped.

Instantaneous systems Where it is not feasible to have a cold water storage tank or hot cylinder (typically in flats), some form of instantaneous water heating is usually provided. This may be a gas-fired multi-point water heater (supplying hot water only) or a combination boiler which also supplies the central heating.

Heating Systems

Mention has already been made of boilers supplying central heating as well as hot water, and the vast majority of homes with central heating have this type of system. The circulation from boiler to radiators is driven by a pump, and on more sophisticated systems, motorized valves control the distribution of hot water to the hot cylinder or the radiators as directed by the system controls (see page 92).

Waste Water Systems

Getting rid of all the waste water a house produces is generally much simpler than supplying it in the first place since you can rely on gravity to do the job for you.

Two-pipe waste system These are generally found in older homes. Here the WC waste runs to a large-diameter soil pipe which runs down the outside wall of the house and which is connected directly to the underground drains at a manhole or inspection chamber. The soil pipe also runs up to roof level, where it is vented to the open air.

Waste water from other upstairs appliances runs to a hopper mounted on the outside of the house wall. It then goes via a downpipe to a gully at ground level and on to the drains via a separate underground pipe. Waste water from downstairs goes direct to an open gully and then via another branch to the drains. Rain-water downpipes often also discharge into the same drains.

Single-stack waste system These are found in newer homes. Here a single soil pipe collects waste from all the water-using appliances in the house and takes it direct to the underground drains. This soil pipe is again usually vented to the open air at roof level, but may be capped inside the house with a special pressure relief valve in certain circumstances. There may also be gullies at ground level to take waste from appliances remote from the soil stack, but the waste pipes discharge into them below the level of a gully grid instead of over it. Rain-water downpipes are no longer connected to the foul-water drains, but run either to separate surface-water drains or to soak-aways.

The twin advantages of this single-stack system are that all the pipework can be concealed within the house, and that it does away with open discharges at gullies and hoppers which lead to smells and blockages. However, they must be designed carefully if they are to work properly, and there are strict rules about where connections can be made into them.

The most important feature of any waste water system is that it must not allow drain smells to enter the house through the waste pipes. All waste systems do this by having water-filled traps connected to every water-using appliance (or built in in the case of WC pans), and these are designed so that they retain the water seal as the appliance is emptied. On modern single-stack systems the trap should always be the deep-seal type, with a water depth of 75mm (3in).

The soil pipe and any branch pipes from gullies run underground from the house and link up at a manhole. This will

Fig 5 (*right*) A modern single-stack waste and soil pipe system, with appliance waste pipes connected directly to the stack. Remote appliances discharge into back-inlet gullies, and the underground pipes connect at inspection chambers or manholes. Rain-water downpipes run to a separate surface-water drain, or into soak-aways.

Plumbing, Heating and Waste Systems

be a brick-built chamber on older properties, or a smaller plastic one on newer ones. From there a single drain carries the combined wastes to the main sewer, which generally runs beneath the road next to the property. Additional manholes are provided wherever the drain run changes direction, and on older properties the final manhole may contain a trap and a capped-off rodding eye – this is known as an interceptor trap, and is designed to prevent sewer gases and rats from entering the system. Such a trap is not needed on modern drainage systems.

A few rural properties remote from mains drainage may have a cesspool or septic tank to dispose of waste water. A cesspool is just a watertight underground chamber which retains sewage and waste water until it can be pumped out and disposed of by a special tanker – a job that has to be carried out regularly. A septic tank is in effect a miniature sewage works, encouraging the decomposition of stored sewage through the action of anaerobic bacteria so that harmless effluent can then be discharged safely into a stream, ditch or land drain.

Fig 6 (*below*) Older properties have a two-pipe waste and soil system, with the WC discharging into a separate soil pipe and waste from appliances going to hoppers upstairs and gullies at ground level. Rain-water downpipes may also discharge into the same drain via gullies.

Fig 7 (*bottom*) An interceptor trap and rodding eye, often found on old properties.

Fig 5

Fig 6

Fig 7

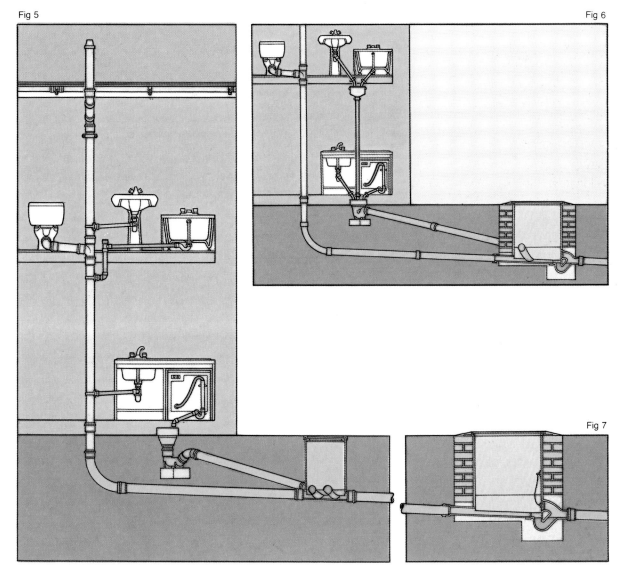

11

Raw Materials

The key materials for any plumbing project are the pipework that carries the water to and fro, and the fittings that connect everything together. Traditionally both have been made of metal, but plastics are becoming increasingly common and have made a virtually complete take-over in the field of soil and waste piping.

Types of Pipes

Copper is still by far the most widely used material for all water supply and central heating pipework in the home. It is strong yet comparatively lightweight, easy to work and relatively inexpensive. The grade used for domestic projects is known as 'Table X', and you are likely to use three common sizes with outside diameters of 15mm, 22mm and 28mm. These are almost identical to the old imperial sizes – ½in, ¾in and 1in, measured as internal diameters – which they replaced (see page 27 for details of how to join new metric pipework to old imperial sizes).

You can also buy short lengths of hand-

Fig 8   Types of pipe.

1 uPVC soil pipe
2 overflow pipe
3 waste pipes
4 polybutylene pipe
5 cPVC pipe
6 polythene pipe
7 rain-water pipes
8 copper pipe
9 stainless steel pipe
10 hand-bendable pipe

bendable copper pipe with corrugated walls which is designed mainly for use in connecting new tap tails without the need to alter existing pipework.

Plastic piping rules the roost when it comes to getting rid of waste water, but until recently it has not been used for hot water supply or heating pipework. This is because the materials tended to soften slightly and sag at higher temperatures, so its use has been restricted to cold water supplies only. Polyethylene is the most widely used material for cold pipes, and a recently-introduced cross-linked variety called Pipex can be used for pipes carrying hot water too.

Two other plastic pipes – polybutylene and chlorinated polyvinyl chloride (cPVC for short) – now also have approval for general plumbing use from the Water Research Centre (the official body which tests and approves plumbing materials). The main advantage of both Pipex and polybutylene pipe is that they are relatively flexible, so can be bent into gentle curves by hand without the need for special tools or fittings. They can also be connected using proprietary push-fit connectors as an alternative to conventional compression fittings. By contrast, cPVC pipe is rigid, and is designed to be connected with slim solvent-welded fittings. All three types are available in 15mm and 22mm sizes.

The commonest plastic for both waste and soil pipes is unplasticized PVC (uPVC), and this is also widely used for rainwater gutters and downpipes. However, two other plastics – acrylonitrile butadiene styrene (ABS) and polypropylene – are also used for waste pipes and, in the case of polypropylene, overflow pipes too. ABS and uPVC pipes are joined by solvent-welded fittings, while polypropylene pipes use push-fit connectors. Standard sizes for all three types are 32mm, 40mm and 50mm for waste pipes, 110mm for soil pipes, 22mm for overflows and 50, 65 or 110mm for downpipes. Note that the pipe and fittings are not always compatible between brands, so care must be taken when mixing components.

On an old plumbing system you may come across iron or lead piping. Lead pipes should be replaced by copper whenever possible, especially in soft water areas, while work on iron piping is generally best left to a plumber.

Plumbing Fittings

The pipe you are working with dictates the type of fitting you use to connect it, although in some situations you have a choice. Here are your options:

Copper pipe Capillary, compression or push-fit fittings.
Polybutylene and **Pipex pipe** Push-fit or compression fittings.
cPVC pipe Solvent-weld fittings.
uPVC pipe Solvent-weld fittings (or push-fit fittings for soil pipes).
ABS pipe Solvent-weld fittings.
Polypropylene pipe Push-fit fittings.

Each is available in a wide range of different forms (see page 14).

Capillary fittings These are made from copper, and the type used for most do-it-yourself plumbing work is the so-called *Yorkshire* fitting. Each end of the fitting contains a ring of solder which you melt with a blowlamp or hot air gun to bond the pipe in place. The solder is lead-based, but the current water by-laws require special fittings with lead-free silver-based solder to be used on pipes carrying drinking water. The main advantages of capillary fittings are that they are cheap and unobtrusive in use.

Compression fittings These generally have brass bodies, with a brass cap nut on each end which retains a soft brass or copper sealing ring called an olive. To make a joint you slip the cap nut and olive over the pipe end, insert it into the fitting and tighten the nut to compress the olive and make a watertight seal. The advantage of compression fittings is that they are easy to assemble and undo, but they are more expensive than capillary fittings and much more obtrusive on exposed pipework.

Plastic compression fittings These are used on some plastic waste systems and work in a similar way to compression fittings, compressing a flexible rubber or plastic sealing ring on to the pipework.

Push-fit fittings These actually come in two varieties – one for supply pipework and the other for waste and soil pipes. The former looks like a plastic compression fitting, but the cap nut retains a steel grab ring that locks the pipe within the fitting, plus a rubber washer and a rigid plastic sealing ring which make the joint

CHECK
- that any copper pipe you buy is made to British Standard BS2871
- that metallic fittings comply with BS864

watertight. To assemble a joint you simply push the pipe into the fitting. They are very easy to use, but are comparatively bulky and expensive.

Push-fit fittings for waste pipes contain a flexible sealing ring only, since they do not have to cope with water at high pressure. To make the joint you simply push the pipe into the fitting, then withdraw it slightly to leave an expansion gap. They are cheap and easy to use.

Solvent-weld fittings These are bonded to their pipework with special solvent-weld cement. They are sleeker than push-fit fittings in appearance and joints are easy to make but are permanent once assembled.

Taps and Valves

The other plumbing components you are likely to use are taps and valves to control the water flow.

Taps come in a wide range of types, and your choice will depend mainly on what you want it to do and the style. (See pages 36 and 37 for more details).

Valves allow you to isolate parts of you plumbing system for repairs or alterations, and come in several types (see page 32). Ball-valves are fitted to water storage cisterns and open and close automatically to refill the cistern when water is drawn off from it (see pages 46, 72 and 74 for more details).

TIP
When buying taps, watch out for Continental imports which are designed for use on high-pressure water systems and which will give poor flow rates if fitted to a traditional low-pressure supply.

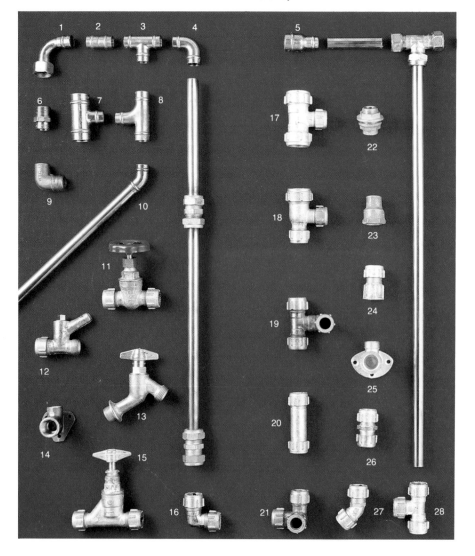

Fig 9 (left) Capillary fittings (1–10) and brass compression fittings (11–28).

capillary fittings
1 15mm angled tap connector
2 15mm straight coupling
3 15mm tee
4 15mm elbow
5 22 × 15mm reducing coupling
6 15mm copper × male iron coupling
7 22 × 22 × 15mm tee
8 15mm swept tee
9 15mm elbow copper × male iron
10 135° elbow

compression fittings
11 gate valve
12 draincock
13 angled bib tap
14 back-plate elbow
15 stoptap
16 15mm elbow
17 22 × 22 × 15mm tee
18 22 × 15 × 15mm tee
19 15mm offset tee
20 extended coupling
21 15mm corner tee
22 tank connector
23, 24 copper × female iron couplings
25 Back-plate elbow
26 15mm straight coupling
27 135° elbow
28 15mm tee

Raw Materials

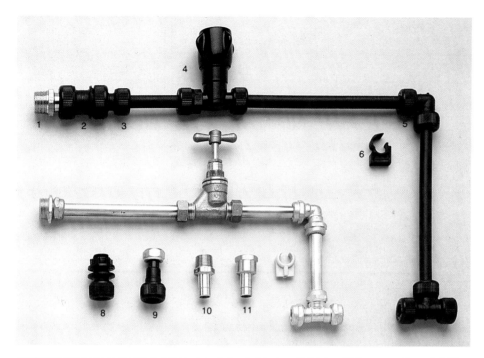

Fig 10 (*left*)   Polybutylene pipe and push-fit fittings, laid out alongside comparable pipe run in copper pipe with brass compression fittings.

1 coupling × male iron
2 22mm straight coupling
3 22 × 15mm reducing coupling
4 stoptap
5 15mm elbow
6 pipe clip
7 15mm tee
8 tank connector
9 tap connector
10 coupling × male iron
11 coupling × female iron

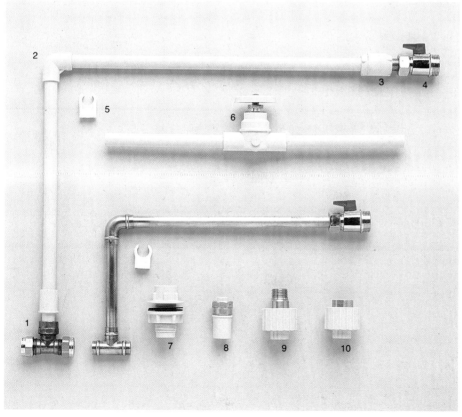

Fig 11 (*left*)   cPVC pipe and solvent-weld fittings, with copper pipe and capillary fittings for comparison.

1 15mm tee
2 15mm elbow
3 coupling with spigot for connection to compression fittings
4 washing machine valve
5 pipe clip
6 stoptap
7 tank connector
8 tap connector
9 coupling × male iron
10 coupling × female iron

Tools and Equipment

For many plumbing jobs, you will only need the general-purpose tools already in your tool-kit – things like screwdrivers, adjustable spanners and hacksaws, for example. You will, however, find it difficult to get by without investing in some of the specialist tools of the plumber's trade as well. It is also a good idea to keep a small, separate took-kit for coping with plumbing emergencies like burst pipes. Below are listed some of the things you will need.

Specialist Tools

The essential plumbing tools you will need are mainly those designed for cutting, bending and joining pipework. There are also one or two items you may need once in a lifetime; it's often cheaper to hire them as and when they are required.

Pipe cutters You can always use a hacksaw for cutting pipework to length, but you will get better results more quickly if you use a pipe cutter with copper pipe. This guarantees a clean, square cut and avoids the creation of metal filings which can cause corrosion within the pipework. Most pipe cutters have a pointed reamer at the end which is used to remove the slight burr which the tool's cutting wheels leave on the inside of the pipe. Choose one which will cope with pipe sizes up to 28mm.

Bending springs It is physically possible to bend copper pipe by hand, but the pipe walls will kink and split at the bend unless they are supported. This support is provided by a bending spring, a strong steel helix which you slide into the pipe and position at the bending point before pulling the pipe round your knee to the angle you require. You then draw the spring out again by means of a length of strong string tied to the loop at the end of the spring. You need one spring for 15mm pipe, another for 22mm pipe; for 28mm pipe, you have to use a pipe bending machine.

Pipe bending machine This is an ingenious tool, consisting of a pair of levers which draw a straight former over a curved one to produce pipe bends of any angle. The formers support the pipe walls as the bend is made, and by changing the former the machine can cope with two or three different pipe sizes. Buy one if you plan extensive plumbing work, or hire one if you need it only occasionally.

Spanners You may have one or more adjustable spanners in your tool-kit already. If you have not, you will need a pair of them to cope with compression fittings and all sorts of other screwed connections. There are several patterns available; probably the best type for plumbing work is the crescent-pattern spanner, which has slim-angled jaws capable of reaching into tight corners. Make sure you select ones which will cope with nuts up to 25mm across.

You will also need a specialist spanner called a *crowsfoot* or *bath-and-basin*

Tools and Equipment

spanner for fitting and disconnecting pipes to taps in the confined space behind most appliances. This has angled jaws at each end, one sized for ½in basin and sink taps, the other for ¾in bath taps. There is a combination version of this called the Taptool, which also acts as a general-purpose spanner.

Lastly, once in a blue moon you may need an *immersion heater spanner* designed as its name implies for screwing an immersion heater into the flange in the wall of a hot cylinder. It's best to hire one when you need it.

Heating tools   If you intend to use cap-illary joints on your pipework, you will need a tool that will heat up and melt the solder inside the fittings.

The *blowlamp* is the traditional tool for this, but many a fire has been caused by a carelessly wielded lamp and the amateur plumber will often prefer an alternative method. The first option is to use a *hot air gun* with a special spoon reflector nozzle to concentrate the air stream on the fitting being soldered. Most are electric, but there are one or two cordless types which run on a gas cartridge like a blowlamp – convenient for work away from a power supply. The second option is to use electric

Fig 12 (*left*) You'll need both specialist and general-purpose tools for most plumbing jobs.

1 crowsfoot spanner
2 pipe bending machine
3 tape measure
4 pipe cutter
5 bending spring
6 half-round file
7 screwdrivers
8 heatproof gloves
9 hacksaw plus spare blades
10 self-grip wrench
11 junior hacksaw
12 pliers
13 pipe wrench
14 adjustable spanners
15 flux paste
16 wire solder
17 jointing compound
18 wire wool
19 blowlamp, hot air gun
    or electric tongs

Tools and Equipment

pipe tongs which are a variation on the soldering iron principle: you simply clamp the tongs round the fitting and press the trigger.

Hole cutters If you need to cut fairly large holes in a replacement storage cistern, the tool to use is either a tank cutter or a hole saw. Both fit in your electric drill; the former is adjustable, while the latter comes in a range of sizes.

Sundry tools There are a few other specialist tools you need in your plumbing tool kit.

The first is a *radiator key*, a small socket spanner used for letting air out of radiators. It's worth having several of these so you can always find one when you need it. Next comes the *plunger* or force cup, essential for shifting blockages in waste pipes and WCs. Lastly, you may need a *tap reseating tool* for recutting worn washer seatings inside taps. Hire one when you need it.

Plumbing Sundries

You'll need flux and possibly some wire solder for making capillary joints. This is electrician's solder, not plumber's solder which is used only for making joints in lead piping. Make sure you have lead-free solder for drinking water supply pipes.

For threaded joints you can use either jointing compound or PTFE tape (short for polytetrafluoroethylene) to ensure a watertight joint. Again, make sure you use suitable compound such as Boss Blue on drinking water supply pipes.

Penetrating oil is useful for loosening stubborn nuts and fittings; an aerosol type is easier to use than liquid oil. Lastly, you will need wire wool for cleaning pipe ends ready for soldering.

General-Purpose Tools

As already mentioned, you will need many of the general-purpose tools in your tool-kit for plumbing work.

An electric drill is essential for jobs such as making holes in joists and fixing things to wall and floor surfaces. Include a cold chisel and club hammer for making large holes through walls and for cutting pipe chases. You will need a saw for cutting the tongues off floorboards – a circular saw or a special floorboard saw are both ideal – and a lever for prising them up

(use your cold chisel at a pinch). You will also need a fine-toothed saw for cutting plastic pipe, a tape measure, pliers, files and assorted screwdrivers. Last but by no means least, a portable work-bench such as a Workmate is invaluable for plumbing work, both as a handy working surface and as a vice.

An Emergency Tool-Kit

Even if you prefer to leave plumbing work to the experts, it is well worth having a small tool-kit to hand so you can cope with plumbing emergencies. Your kit should include the following tools: an adjustable spanner; a self-grip wrench (which will do double duty as a second spanner); a pair of pliers; a junior hacksaw; a file; a couple of screwdrivers; and a plunger.

Add to these one or two compression fittings in 15, 22 and 28mm sizes, plus a few short lengths of pipe and spare olives in these sizes so you can replace leaky joints or make permanent repairs to burst pipes. It's also worth keeping a selection of spare tap washers and O-rings for repairs to dripping taps. Add a roll of PTFE tape and some penetrating oil too.

For emergency repairs to burst pipes, include a proprietary pipe clamp or a pack of two-part epoxy repair putty.

Fig 13 (*above*) Assemble a small 'first aid' tool-kit and keep it inside the house so you can cope with plumbing emergencies without delay. It should include an adjustable spanner, a self-grip wrench, pliers, a junior hacksaw, a couple of screwdrivers, a plunger, some short lengths of copper pipe, some compression fittings plus spare olives, tap washers and O-rings, penetrating oil, PTFE tape and a pipe clamp or two-part epoxy repair putty for fixing bursts and leaks.

Where best to shop for plumbing fittings and equipment depends largely on the scale of your operations. If all you want is a tap washer or a sink plunger, look no further than your local hardware store, but if you are working on a larger scale you will make considerable savings on your final bill by shopping around. Here are the places to try.

Local DIY Shops

The typical independent high street DIY shop usually stocks a small selection of plumbing goods. You will probably be offered a choice of just one brand, probably blister-packed and in a range extending only to obvious things like common pipe fittings and repair kits. It may also stock small amounts of pipe, which will be sold by the metre – rather expensively. It will probably not have any specialist tools.

Verdict   Fine for small jobs if you do not mind paying more than you need to for the convenience of shopping locally; expensive for larger projects.

DIY Superstores

The major national chains generally offer a good range of plumbing fittings, tools, materials and sundries, although you may be restricted in choice of brand. Some offer parallel ranges of branded and own-brand accessories, often blister-packed. One or two chains also stock heating equipment, and offer useful advice leaflets on simple plumbing jobs.

Verdict   A good selection of tools and materials at fairly reasonable prices, but unlikely to offer much technical advice.

Specialist Retailers

There is a growing number of specialist bathroom and kitchen retailers around, offering an excellent range of appliances, and many are also happy to offer sound and useful technical advice. You should be able to buy everything you need for even the most complex kitchen or bathroom plumbing project, but they rarely stock tools and may not have a wide range of fittings.

Verdict   Excellent selection of equipment, generally at reasonable prices; good for technical advice too.

Builders' Merchants

Most usually stock a reasonable range of plumbing and heating fittings, accessories, sundries and tools, but may be loyal to only one or two brands. Some have 'retail' counters designed to cater for the non-trade customer.

Verdict   Reasonable selection of goods at a reasonable price.

Plumbers' Merchants

Plumbers' merchants stock by far the widest range of plumbing tools, fittings, accessories and other equipment, often offering several different brands. However, since they exist to cater for the professional plumber, you need to know what you want before you start – often very little is on display. It is worth contacting local firms for estimates if you plan any large-scale projects.

Verdict   By far the biggest choice of materials and tools, generally at the keenest prices; good for large orders.

Mail Order Suppliers

There are several firms operating a mail order delivery service for plumbing goods, especially central heating equipment. They publish comprehensive catalogues and usually offer a wide range of brands. Most also offer a complete system design service based on your room measurements and temperature requirements.

Verdict   Good for plumbing and central heating materials only, often at surprisingly keen prices.

Buying in Bulk

If you are planning extensive work on your plumbing system, you will make considerable savings by buying basic things like pipe and fittings in bulk. Pipe is usually sold in bundles of 5 or 10 lengths (usually 3m or 10ft long), while fittings are sold in 10s, 20s, and 50s.

SAFETY
There is rarely any actual danger inherent in plumbing work itself, but accidents can happen because of misuse of tools or equipment. Points to watch out for include:

- fire risks, especially with blowlamps – always keep a fire extinguisher handy when using one
- fumes in confined spaces – always make sure work areas are well ventilated
- electrical hazards – take care not to drill or cut through wiring in walls or under floorboards. Use a plug or adaptor fitted with a residual current device (RCD) to connect power tools to the mains
- access equipment – make sure ladders, work platforms and steps are standing squarely and are secured if necessary
- gas – always call in a professional gas installer to carry out any work on your gas supply. See page 94 for useful addresses

Cutting Copper Pipe

Despite the growing use of plastic piping for water supplies and heating circuits, copper remains by far the most widely used material for domestic plumbing work. Even if you decide to use plastic for new pipework in your home, you will need to know how to work with copper pipe when altering or repairing your existing plumbing and heating pipework which is certain to be run in copper pipe.

There are two situations in which you will need to cut copper pipe. The first is when you are carrying out new work and have to cut individual pieces to the length you require. The second is when you are making new connections to existing pipework – (see page 27). In both cases your aim must be to produce a clean, square cut without deforming the pipe cross-section or leaving ragged swarf along the line of the cut. If you do not take care over this, the pipe ends will not fit properly within the connecting pipe fittings and leaks may result, while swarf can lead to corrosion in the system.

What to do

When cutting 'loose' pipe you can use either a hacksaw or a pipe cutter. If you use a hacksaw, mark the cutting position on the pipe and hold it securely (without squashing it) in a vice or the jaws of a Workmate. Draw the saw blade lightly over the pipe to start the cut, and then saw through it slowly and carefully so the cut is square. Complete the cut and file off any rough edges.

A pipe cutter is quicker and more accurate than a hacksaw, guaranteeing a square cut without swarf, but it cannot be used to cut into existing pipework. To use it you simply fit the pipe in the cutter's jaws, tighten the knob and rotate the cutter round the pipe so the wheels cut through the pipe wall. This leaves the pipe wall bent inwards very slightly; you then use the reamer at the end of the cutter to straighten it again. The standard type will cut pipes up to 25mm in diameter; you'll have to use a hacksaw on larger sizes.

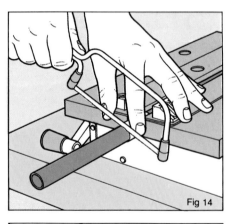

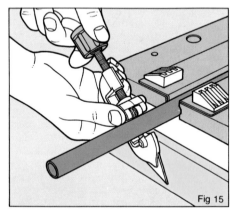

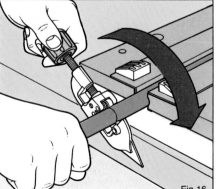

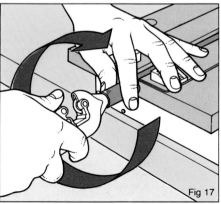

What you need:
- junior hacksaw
- spare hacksaw blades
- vice or Workmate
- steel tape measure
- flat file
- pipe cutter

CHECK
- when using a hacksaw that the cut is square, and that any swarf is filed away before the pipe is connected up
- when using pipe cutters that the pipe end is reamed out after cutting

TIP
If you plan to use soldered joints, polish the last inch or so of the pipe ends with wire wool to ensure that the solder forms a good bond with clean metal.

Fig 14   To cut pipe with a hacksaw, hold it in a vice and saw steadily down at the cutting mark. Finish the cut carefully so you avoid leaving ragged swarf.

Fig 15   With a pipe cutter, fit the jaws round the pipe with the cutting wheel in line with the mark, and tighten the end knob.

Fig 16   Rotate the cutter round the pipe. Tighten the jaws very slightly as the cut is made.

Fig 17   Ream out the cut end of the pipe using the V-shaped bar at the nose of the tool.

Cutting Plastic Pipe

Copper may still be the commonest choice for supply pipework, but plastic pipes are now used universally for waste and soil pipes. They are also becoming increasingly popular, especially with do-it-yourself plumbers, for all new supply and waste pipework within the home because of their ease of use.

The smaller diameters are used for waste pipes running from appliances such as baths, basins and sinks, and also for overflow pipes from storage and WC cisterns. Larger sizes are used for soil pipes from WCs and for the vertical stacks that carry soil and waste water to the underground drains.

As with copper pipe, it is important to cut plastic piping carefully so that the ends are square and free from swarf and will seat properly in the fittings used to connect lengths together. However, the material is much easier to cut than copper and there is not the same risk of deforming it as you do so.

What to do

You can cut all types of plastic piping with any fine-toothed saw; a tenon saw is ideal for smaller sizes because its wide blade makes it easier to ensure a square cut, but you can use a hacksaw instead, or even an electric jigsaw. A panel saw is a better bet for large-diameter soil and drainage pipes.

To guide the cut, wrap masking tape or a piece of paper round the pipe in line with the cutting mark and hold it securely in a vice or Workmate. When the cut is almost complete, rotate the pipe so you can finish the cut from the opposite side. Then use abrasive paper or wire wool to remove any plastic swarf and to leave the pipe end perfectly smooth.

The makers of polybutylene pipe recommend the use of their own special pipe secateurs, although a fine-toothed saw or serrated kitchen knife can be used just as effectively.

What you need:
- fine-toothed saw
- vice or Workmate
- steel tape measure
- secateurs for polybutylene pipe
- abrasive paper or wire wool

CHECK
- that cuts are square and swarf is removed before joints are made

TIP
Polybutylene pipe has V-shaped marks on it at regular intervals. Cut at one, and the next acts as a handy guide to how far the pipe should be inserted into its push-fit connector.

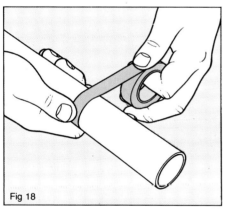

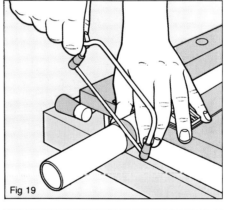

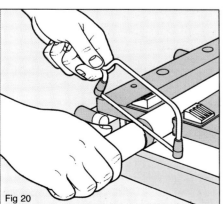

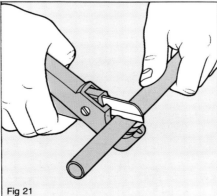

Fig 18   To ensure a square cut, wrap masking tape or a piece of paper round the pipe at the cutting mark.

Fig 19   Hold the pipe in a vice or the jaws of a Workmate and saw down at the cutting mark.

Fig 20   When the cut is almost complete, rotate the pipe and finish the cut from the other side.

Fig 21   With polybutylene pipe you can use the maker's special pipe secateurs if you prefer. They work just like gardeners' secateurs.

Bending Copper Pipe

You could plumb your entire house without ever bending a length of pipe by using elbow fittings to turn all the corners. However, you would not get a very neat or attractive installation, and you would spend a lot of money unnecessarily. Bends are free; all you have to do is learn how to make them.

The key thing to remember is that while copper pipe can be bent by hand, it will kink and split if you simply grab the pipe ends and pull. The walls of the pipe need supporting while the bend is made so that the metal will deform evenly and without damage.

A pipe bending spring is the tool to use. It is a helix of hardened steel which you insert into the pipe, centre on the bend position and draw out again after the bend is made. You need a separate spring for each size of pipe you are working with.

However, if you are planning extensive plumbing work, it is worth buying a pipe bending machine; you can always sell it afterwards if you then have no further need for it.

What to do

If you are using a bending spring, choose one to match the pipe diameter and oil it lightly. Tie a length of string to one end so you can retrieve it from the pipe when you have finished. Lay the spring alongside the pipe to be bent so you can mark the position of the end of the pipe on the string as a guide to placing the spring correctly. Slide the spring into the pipe, and form the bend by pulling the pipe round your knee. Overbend it slightly, then pull it back to the required angle – this helps to release the spring, which can then be drawn out easily.

With a pipe bending machine, mark the position of the crown of the bend on the pipe, place the pipe in the former and fit the backslide. Line up the mark on the pipe with the former using a pipe offcut, then pull on the lever to bend the pipe round the former. When you have bent the pipe to the required angle, open the levers and remove the backslide to release the pipe.

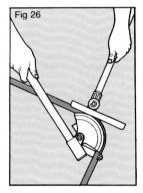

Fig 26   A pipe bending machine draws a straight former over a curved one to produce bends of any angle.

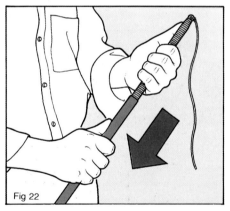

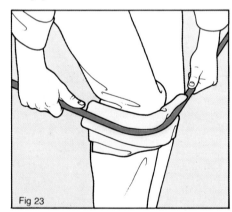

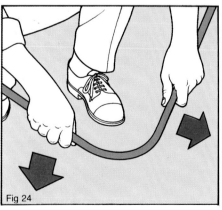

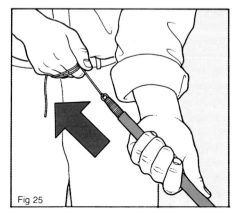

Fig 22   Slide the lubricated spring into the pipe.

Fig 23   Bend the pipe over your knee, using a pad of cloth as protection.

Fig 24   Bend the pipe back to the required angle.

Fig 25   Finally, draw out the spring.

Using Compression Fittings

Of the various ways available for making joints in copper pipe, the amateur plumber usually turns to compression fittings as a first choice – presumably because they are easy to use, needing just a couple of spanners. However, they are also quite expensive, and do not look very neat on exposed pipework; capillary fittings (see page 26 for more details) are much less obtrusive in use.

Compression fittings are generally made of brass and can be used with stainless steel and some types of plastic pipework as well as with copper. With plastic pipes, special metal sleeves have to be inserted into the pipe ends first so the olive is not distorted as the cap nuts are tightened. They work by compressing a deformable metal ring called an *olive* on to the pipe, and then drawing this tightly against the body of the fitting to make a watertight seal. The fitting can easily be dismantled if required – simply loosen the cap nuts to release the pipe.

What to do

Start by cutting the pipes to be joined, and check that the ends are square and free from swarf or burrs. Then slip a cap nut and an olive over each pipe end and push the first pipe into the fitting so it butts up against the internal depth stop. Push the olive against the mouth of the fitting, slide up the cap nut and tighten it by hand as far as you can. Grip the body of the fitting with one spanner and the cap nut with another, and tighten up by 1¼ turns for 15mm pipe, 1 turn for 22mm pipe and ¾ of a turn for 28mm pipe. If you over-tighten the nuts you will deform the olive and the fitting will leak. Repeat this for the other pipe(s) entering the fitting to complete the joint.

You should not need any sealant to make a new compression fitting watertight, but on a re-made fitting it can help to wrap a little PTFE tape over the face of the olive next to the fitting.

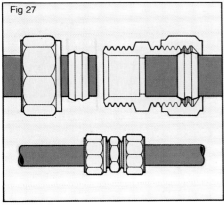

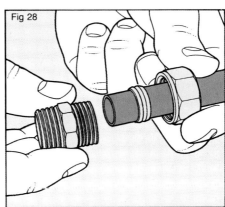

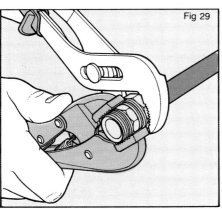

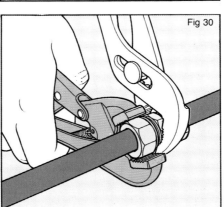

Fig 27   At each end of a compression fitting, the pipe is held in place by a threaded cap nut which also deforms the olive on to the pipe as it is tightened.

Fig 28   Place a cap nut and olive over the pipe end, and push the pipe into the fitting until it meets the depth stop.

Fig 29   Push up the olive against the mouth of the fitting and hand-tighten the cap nut. Then grip the fitting with one spanner and tighten the cap nut with another.

Fig 30   Repeat the process to connect the other pipe(s) to the fitting.

23

Using Push-Fit Fittings

Plastic push-fit joints were developed primarily for use with polybutylene pipe, and they can also be used on copper or stainless steel pipe and are suitable for both water supply and heating pipework. Their main advantages are that they can be assembled and undone by hand (no tools are needed at all, which can be a boon if you are trying to make a connection in a confined space) and the joint can be rotated without affecting the watertight properties of the seal. Set against these advantages are their expense, their relatively bulky appearance and the limitation that their performance is not guaranteed at temperatures above 90°C (194°F) so they cannot be used for connections to a boiler, or within 1m (3ft) of one.

The fitting works by locking a toothed steel grab ring on to the pipe, preventing it from pulling out of the fitting, and by compressing a plastic seal and a rubber O-ring against the body of the fitting to make a watertight joint.

Push-fit joints of a slightly different type are also used with some waste pipe systems (see Fig 34). Here internal O-rings in the fitting provide the watertight seal.

What to do

Start by cutting the pipe to leave a neat square end. With polybutylene pipe, make the cut at one of the V-shaped marks stamped on the pipe so you can use the next mark as a guide to inserting the pipe to the correct depth in its fitting. Then insert a steel sleeve into the pipe end (polybutylene and polythene pipe only) to prevent the pipe being deformed.

Don't dismantle the fitting to assemble the joint; all you have to do is to push the pipe ends fully into the fitting and then withdraw them slightly to ensure that the grab ring has locked on to the pipe.

If you have to dismantle a push-fit joint, you must always prise off the old grab ring with pliers and fit a new one before reassembling the joint.

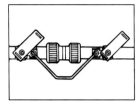

Fig 35.

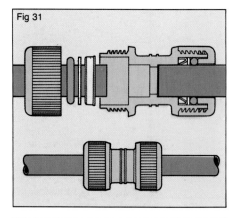

Fig 31

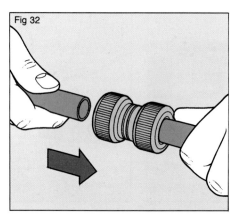

Fig 32

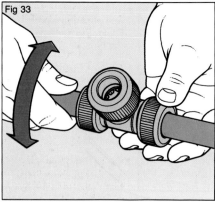

Fig 33

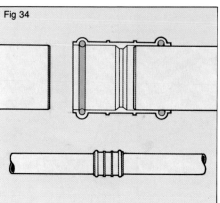

Fig 34

Fig 31 At each end of a push-fit fitting, the pipe is held in place by a toothed grab ring and a cap nut. The O-ring and plastic seal ensure a watertight joint.

Fig 32 To make the joint, simply push the pipe fully home into the fitting, then withdraw it slightly to ensure that the grab ring has gripped the pipe.

Fig 33 The pipes can be rotated within the fitting without affecting the seal.

Fig 34 Push-fit joints on waste pipe runs contain internal O-rings that grip the pipe as it is pushed home.

Using Solvent-Weld Fittings

Plastic solvent-weld fittings are used to make joints on waste, soil and overflow pipes made from PVC or ABS plastics. The joint is literally glued together using special solvent-weld cement that softens the plastic and dries to form a strong bond that is rigid and permanent. Their main advantage is that they are neater and cheaper than other fittings; their drawbacks are their permanence which makes it difficult to carry out alterations or repairs to the pipe run without major surgery, and the fact that you need to incorporate expansion joints on long pipe runs to prevent distortion caused by the pipe heating up.

The same solvent-weld principle is used on the innovative Hunter Genova small-diameter pipe system, which can be used for water supply and central heating pipework (subject to the same restrictions as polybutylene pipe as far as connection to boilers are concerned). It is made from chlorinated PVC (cPVC).

What to do

There are several makes of push-fit waste fittings on the market, but not all are mutually compatible so it is best to use fittings and pipe that match existing waste pipes for additions and alterations.

Start by cutting the pipes to length. Degrease the pipe end and the spigot ends of the fitting with the manufacturer's special cleaner or fine abrasive paper. Brush the solvent-weld cement on to both pipe and spigot, then push the pipe into place, rotating it slightly to spread the cement evenly within the joint. Leave to set as directed by the manufacturer before running water through the joint.

To allow the pipe to expand safely, incorporate a push-fit expansion joint in runs over 1.8m (6ft) long.

Use the identical jointing procedure for making joints with small-bore cPVC pipe. Special fittings are available for making connections to pipes of other materials.

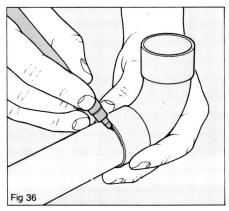

Fig 36

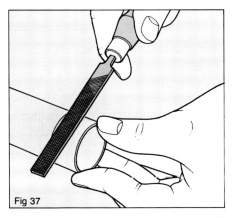

Fig 37

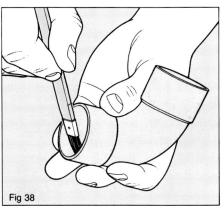

Fig 38

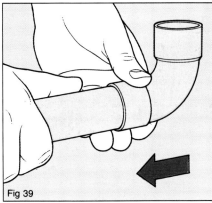

Fig 39

Fig 36 Insert the pipe end in the fitting and mark round it with a pencil.

Fig 37 De-grease the pipe end and the inside of the fitting, and roughen the pipe with a file or abrasive paper.

Fig 38 Brush solvent-weld cement on to the pipe end and the inside of the fitting.

Fig 39 Push the pipe into the fitting, and rotate it slightly to spread the adhesive evenly.

Using Capillary Fittings

Capillary fittings are the first choice of the professional plumber for several reasons. They are inexpensive compared to other fittings, joints are quick and easy to make and assembled pipe runs look neat and streamlined. Many amateur plumbers are put off using them through a combination of fear and uncertainty: fear of using an unfamiliar technique and uncertainty as to whether the joint will be watertight. Yet all you need is confidence, especially since you do not even have to use a blow-lamp if that is what is putting you off – hot air guns and special soldering irons can be used instead.

Most plumbers use pre-soldered York-shire fittings which contain a ring of solder inside the copper body of the fitting. Heat melts the solder, and it then flows by capillary action into the narrow gap between pipe and fitting to form a watertight joint, setting hard as soon as the heat source is removed. Note that capillary fittings can be used only on dry pipework; any moisture present will cause joint failure.

What to do

You can make soldered joints only on pipework that is empty of water, so they cannot generally be used on existing pipe runs unless these can be completely dried out – use another type of fitting instead.

To make the joint up, start by cleaning inside the mouths of the fitting and the outsides of the pipe ends with steel wool – this will remove any tarnish and ensure a good joint. Then brush a little flux – a paste that helps the solder run freely – inside the fitting and on to the pipe end and assemble the joint. Wipe off any excess flux, then apply heat to the fitting. You will see a complete ring of solder form around the mouth of the fitting; you then move the heat to the other part(s) of the fitting and repeat the process. Do not disturb the joint until it has cooled down.

If you use heated tongs, simply clamp the jaws on each end of the fitting in turn and heat it for a few seconds to melt the solder and make the joint.

What you need:
- capillary fittings to match pipe sizes
- steel wool
- flux
- blowlamp, hot air gun or electric tongs

CHECK
- that pipe ends are pushed into fittings until they meet the internal depth stop
- that walls and other nearby surfaces are protected from heat with a glass-fibre mat if using a blowlamp or hot air gun

LEAD-FREE SOLDER
Fittings with lead-free solder must now always be used on pipework carrying drinking water

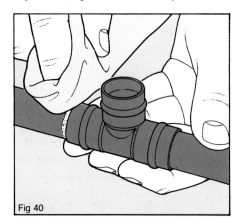

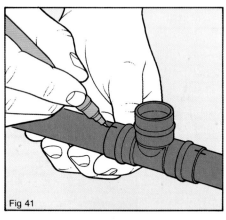

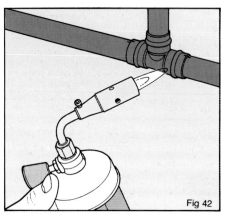

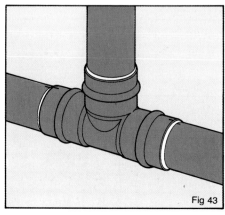

Fig 40   Smear flux on the pipe end and inside the fitting, push the pipe in until it reaches the internal depth stop and wipe off excess flux.

Fig 41   On tees, mark the pipe and fitting so that the branch remains correctly aligned as you make the joint.

Fig 42   Apply heat evenly to each part of the joint in turn using a blowtorch, hot air gun or electric tongs.

Fig 43   Continue heating until a bright ring of solder appears around the mouth of each socket on the fitting.

Connecting into Existing Pipework

Many plumbing projects involve making connections into existing pipework, mainly into supply pipe runs. The principle is basically the same in every case; you need to cut out a short section of the existing pipe and insert a tee to which your branch pipe is then connected. The only complication is caused by the simultaneous existence of imperial and metric copper pipe – this means that metric fittings can be used in some cases while in others adaptors are required to make the connections.

What to do

With copper pipe you can use compression fittings, capillary fittings (if the pipe can be dried out) or push-fit fittings.

With *compression* fittings, hold the tee against the pipe and mark on it the distance between the internal pipe stops. Cut out the section of pipe, put a cap nut and olive on each end and spring the ends into the tee, tighten the cap nuts to complete the job (see Fig 45). You will have no

problems connecting into old ½in or 1in pipe using a 15mm or 28mm metric tee, but with ¾in pipe and a 22mm tee you will need to use ¾in olives on the old pipes.

With *capillary* fittings, metric tees are compatible again with old ½in and 1in pipes, but not with ¾in pipe. Here the best solution is to make up a section of new pipe with a metric tee joined to two metric/imperial adaptors, and insert this in the pipe run (see Figs 46 and 47).

With *push-fit* fittings, follow the same sequence as for compression fittings. You may have to buy a special metric/imperial O-ring kit to make satisfactory connections to old ¾in imperial pipework, but 15mm push-fit tees are compatible with old ½in pipework.

An alternative to cutting in a branch pipe for connections to washing machines and dishwashers is to use a self-cutting connector (see Fig 48) to which the machine hoses are attached (see page 60 for more details).

(see page 60 for more details)

What you need:
- compression, capillary or push-fit tee to match pipe sizes
- ¾in olives
- 22mm (¾in) capillary adaptor
- tape measure
- hacksaw
- spanners or blow lamp
- branch pipe

TIP
If you are using compression fittings on a vertical pipe run, use a clothes peg to retain the lower cap nut and an olive on the pipe while you assemble the joint. (See Fig 44.)

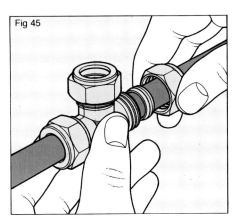

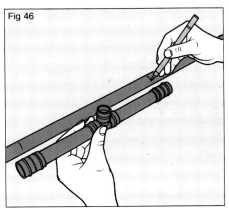

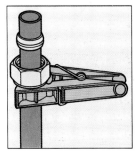

Fig 44.

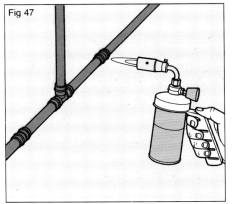

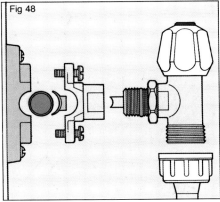

Fig 45 With compression fittings, cut out a short length of pipe and spring the pipe ends into the fitting.

Fig 46 With capillary connections to imperial pipes, make up an assembly of metric tee and two adaptors and mark its length on the pipe run.

Fig 47 Spring the assembly into place and solder the joints.

Fig 48 Use a self-cutting connector for tees to washing machines and dishwashers.

Making Connections to Appliances

The vast majority of the plumbing connections you make will link lengths of pipe together. However, you will also have to connect your supply pipework up to things like taps, cisterns, boilers, hot cylinders and storage tanks, so you will need to be familiar with the various types of connectors available for these tasks. You may also be faced with making connections to existing pipework runs in materials such as lead and those that are galvanized, now no longer used but still present in many older homes (see Tip).

What to do

For making connections to taps you need a tap connector (see Fig 49). This fits inside the threaded tap tail and is secured to it by a backnut; a fibre washer forms a watertight seal between the connector and the tap. Tap connectors are available as compression, push-fit or capillary fittings, in straight and right-angled versions and in two sizes – ¾in for bath taps, ½in for other taps.

There is also a special connector for wall-mounted garden taps, called a backplate elbow. The supply pipe enters one end of the elbow, while the tap screws into the other and the plate is screwed to the wall.

Some pieces of plumbing equipment, such as hot cylinders, have openings with an internal (female) thread into which pipe fittings with an external (male) thread are screwed to make the connections to the system's pipework (see Fig 50). Similarly, other such as boilers may have projecting male threaded inlets and outlets on to which fittings with a female thread are screwed (see Fig 51). All these fittings are still made in imperial rather than metric sizes.

For making connections to storage tanks and securing ball valves in cisterns, you need fittings with backnuts (see Fig 52). Tank connectors have a flange that fits on the inside of the tank and a projecting male thread on the outside to which the supply pipe is then connected by using a female iron coupling.

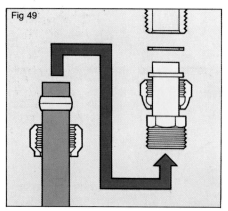

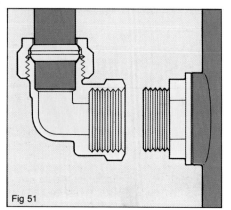

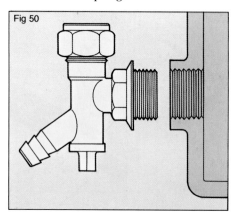

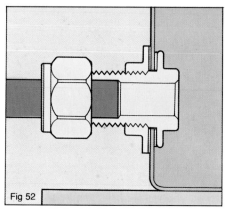

Fig 49 Tap connectors are used to connect supply pipework to tap tails and to ball valves.

Fig 50 Fittings with a male threaded tail are designed for connecting to appliances with an internal female thread.

Fig 51 Where appliances have projecting male threaded inlets or outlets, use a fitting with an internal (female) thread.

Fig 52 Tank connectors link take-off pipes to storage cisterns.

Getting Rid of Waste Water

Water supply is only half the plumbing story; what goes in must come out, and getting rid of water from appliances via the drains is just as important a part of the system.

All water-using appliances are connected to the drains via individual waste pipes. In older homes, appliances like baths and sinks discharged their waste water into hoppers upstairs and gullies downstairs, each individually linked to the underground drains, while WCs discharged into a separate soil pipe. In modern homes, all waste water is discharged via a single soil stack, with the waste pipes from individual appliances connected directly into the stack.

Waste pipe runs and soil pipes are almost exclusively in plastic nowadays, using either push-fit or solvent-welded joints. Making additions and alterations to the system therefore is largely a matter of choosing components compatible with what exists already.

What to do

If you have a system with hoppers and gullies, there is no reason why you should not run extra waste pipes to discharge into them, although in the case of hoppers alternative arrangements are preferable if they are possible. As far as discharges into gullies are concerned, waste pipes should always discharge below the level of the gully grid to avoid the risk of a grid blockage causing a messy overflow.

With modern soil stacks, new waste pipes can be connected directly into the stack, either by solvent-welding the pipe into a spare boss on the stack if one is available, or else by attaching a strap-on or solvent-welded boss to a hole bored in the wall of the stack. Care must be taken to avoid cross-flow between wastes attached to the stack at the same level, and there are other restrictions that govern where additional waste pipes may be connected to the stack (*see* Check).

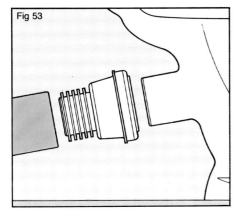

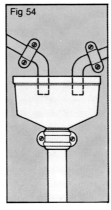

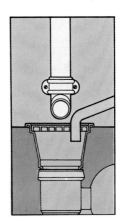

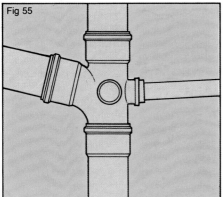

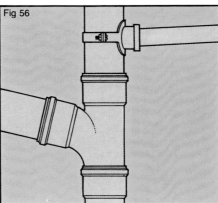

Fig 53 Modern push-fit connectors make it easy to link new WCs to all types of soil pipe.

Fig 54 On old systems, run new waste pipes to hoppers or gullies.

Fig 55 On modern stacks, connect pipes to a spare boss if one exists, or . . .

Fig 56 . . . Fit a strap-on or solvent-weld boss to the stack wall.

Fitting Traps

The waste pipe from every water-using appliance must incorporate a trap which reseals itself with water whenever the appliance is emptied – it prevents drain smells, bacteria and insects from getting into the house. WCs have integral traps, but for other appliances the trap is usually fitted immediately beneath the waste outlet. All are designed to be easy to dismantle for clearing blockages.

The commonest types are the P-trap, the S-trap, the bottle trap and the running trap; combination traps are designed to cope with wastes from double sinks, overflows and appliances such as washing machines or dishwashers.

What to do

To fit a trap to an appliance waste outlet, choose a suitable type for the purpose and for the space available. Remember that on modern single-stack systems, deep-seal traps with a trapped water depth of 75mm (3in) must always be used.

Connect the trap to the waste outlet of the appliance so you can measure and cut the lengths of pipe needed to make up the waste run. Then remove the trap, fit it to the waste pipe and reconnect it. The connection between pipe and trap is either a push-fit joint or a compression type incorporating a rubber or plastic washer.

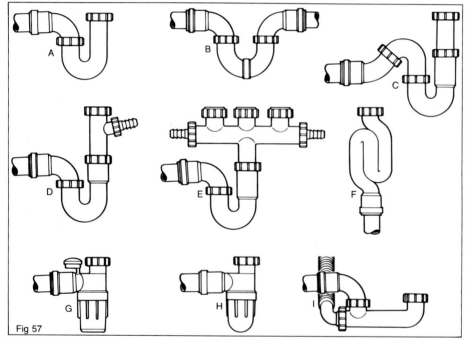

Fig 57

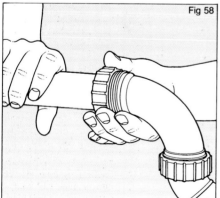

Fig 58

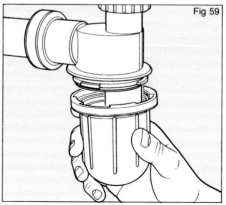

Fig 59

Fig 57 Plastic traps come in several styles and patterns, including: the P-trap (A); running trap (B); universal trap with swivel outlet (C); washing machine trap (D); multiple-inlet trap (E); straight S-trap (F); anti-siphon trap (G); bottle trap (H); and bath trap with overflow (I).

Fig 58 To connect or disconnect a trap from its pipe run, simply twist the cap nut. Make sure sealing washers sit squarely in place.

Fig 59 On bottle traps, unscrew the base to clear blockages.

Running Pipework

Plumbing supply pipework, like house wiring, can be run on the surface but it is preferable to conceal it wherever possible within wall and floor structures. Vertical pipe runs can be plastered over, concealed within stud partition walls or hidden by boxing-in, while the voids beneath suspended timber floors provide the perfect site for horizontal runs. It is also possible to bury pipework within solid concrete floors at the construction stage, but in homes with solid floors it is easier to run new pipework on the surface and to conceal it with skirting conduit round the perimeter of the room. This also makes future maintenance easier to carry out, since the pipes are readily accessible.

What to do

Pipe runs should be as direct and free from bends and elbows as possible, so the first step in planning any plumbing project is to work out the best route for the pipe to take from its source to its destination.

In underfloor voids, clip pipes running parallel to joists, or support them on battens. Set pipes running at right angles to the joists in notches cut in their top edges, but support them on simple hangers where access to the space below permits. Polybutylene pipe is flexible enough for you to feed it through holes drilled in the centre of the joists.

What you need:
- pipe and pipe clips
- plumbing tools
- general-purpose tools

CHECK
- that copper pipes are adequately supported at the following intervals:

	Horiz	Vert
15mm	1.0m	1.5m
22mm	1.8m	2.4m
28mm	2.0m	2.4m

For plastic pipes, space clips 500mm apart on horizontal runs and 1m apart on vertical runs.

TIP
Mark the positions of all pipe runs on floorboards to make them easy to locate in future for alterations or repairs (and to avoid the risk of drilling or driving a nail through them). Label pipes in underfloor voids to identify their purpose.

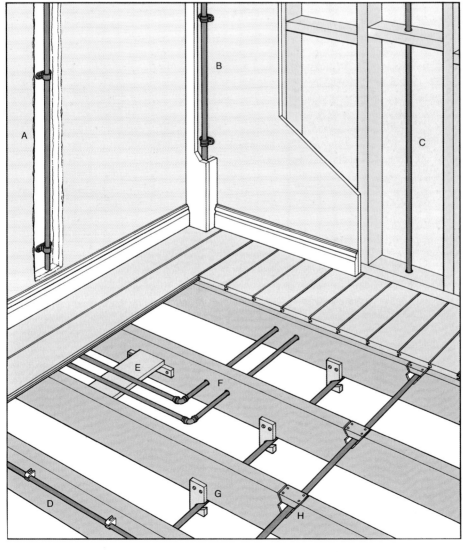

Fig 60   Pipe runs in walls can be buried in plaster (A), boxed in (B) or hidden within partition walls (C). Runs in floor voids parallel to joists should be clipped to the joist sides (D) or laid on battens (E). Runs crossing the joist line can be fed through holes in the joists (F) if the pipe is flexible enough, supported on simple hangers (G) or set in notches in the joist edges (H). In rooms with solid floors, new pipe runs are best hidden in skirting conduit .

Isolating and Draining Pipework

It is vital to know where to turn off the water supply to your plumbing and heating systems, whether you are carrying out a plumbing job or coping with an emergency such as a burst pipe.

Within the house you should have at least one on/off control – a stoptap on the rising main close to where it enters the house. This will shut off the incoming mains-pressure water supply, allowing you to drain pipes, cisterns and storage tanks so you can work on them. Make sure you know where yours is, and show the rest of the family too, so someone can turn it off in an emergency to prevent a flood.

You may also find other controls called gate valves fitted on the outlet pipes from your cold water storage cistern. These allow you to isolate and drain down just part of the system without the need to operate the rising main stoptap. On the best systems, there will also be gate valves on the supply pipes leading to individual appliances such as baths, basins, sinks and WC cisterns, allowing just that tap or ball valve to be isolated.

There will also be draincocks fitted at various points on the system. These allow you to attach a hosepipe and drain water from the pipework or appliance the draincock serves. Likely locations are just above the rising main stoptap, close to the hot cylinder and at one or more low points on the central heating pipework (often close to the boiler).

What to do

How you go about isolating and draining down part or all of your system depends on what controls are fitted. Below are some of the likely options.

Rising main stoptap only If you have no other controls, turn off this stoptap to prevent water entering the house. Drain the rising main by opening the kitchen cold tap so you can work on the mains-pressure pipework. Empty the cold water

Fig 62

Fig 63

Fig 64

Fig 65

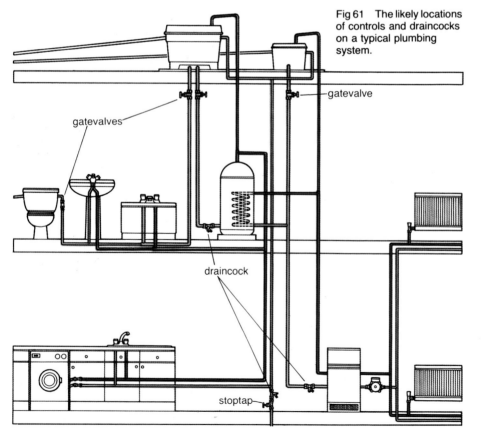

Fig 61 The likely locations of controls and draincocks on a typical plumbing system.

gatevalve

gatevalves

draincock

stoptap

Isolating and Draining Pipework

storage tank by turning on the cold taps in the bathroom so you can work on the low-pressure hot and cold supply pipework (the hot cylinder will stay full if there is no flow into it from the storage tank, and needs draining only for repair work on the cylinder itself or on pipework entering or leaving it).

Gatevalves on cold supply pipes from storage tank If these are fitted, identify which supplies the cold taps and which supplies the hot cylinder, and turn off whichever one is appropriate to allow you to work on the low-pressure supply pipework. Note that gate valves should be fully open when operational. There is no need to turn off the rising main stoptap.

Gatevalves or other isolating taps close to appliances Turn these off to isolate the appliance concerned (*see* Know your Controls on page 32); there is no need to turn off any other valves or stoptaps.

No rising main stoptap In the unlikely event of there being no stoptaps at all anywhere within the house (still the case on some old, unmodernized properties),

you can work on the low-pressure pipework by tying up the arm of the ball valve on the storage tank so the valve stays closed. To isolate the rising main itself, you have no option but to find and operate the water authority stoptap outside. This is usually located beneath an iron cover plate in the pavement in front of the house in a path or drive surface, or occasionally elsewhere in the garden. It is set in a vertical shaft with the tap between 600mm (2ft) and 1m (3ft) below ground level.

To operate the tap you need to check whether it has a conventional T-bar handle or a square metal shank. To turn the former, cut a V-notch in the end of some scrap timber and add a handle so you can grip and turn the tap handle. With the square shank type, you need a special metal key, available from plumbers' merchants. In an emergency you could improvise by flattening the end of some copper pipe slightly so it fits over the shank of the tap. If all else fails, call the water authority's emergency number and ask them to come and turn off the flow for you.

Draining down via a draincock To empty a section of pipework or the hot cylinder, attach a length of ordinary garden hosepipe to the fitting's outlet and lead it to a gully or drain. Then open the drain-cock valve with a small spanner or a special draincock key to allow the water to drain away.

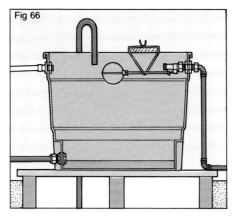

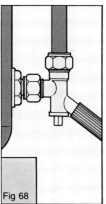

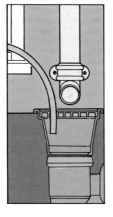

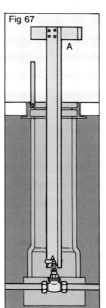

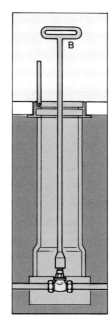

Fig 66 If there is no rising main stoptap (or the tap is jammed), you can still work on the low-pressure pipework. Tie up the storage tank's ball valve arm to stop the flow and open the bathroom cold taps to drain the cistern.

Fig 67 If you need to work on the mains-pressure pipes in the same situation as in Fig 66, you must turn off the outside stoptap. Use a notched wooden key for taps with a T-bar handle (A), and a special key for taps with a square shank (B).

Fig 68 To drain a pipe or appliance, fit garden hose to its outlet, run the hose to a gully and open the valve with a small spanner.

Undoing Things

One of the biggest problems with working on your plumbing system is that you are generally trying to alter, extend or repair what is already there. This means that you are at the mercy of components that may not work as they are supposed to – from the radiator connection that has locked solid to the tap connector that you simply cannot undo. Below are solutions to some of the commonest problems you are likely to encounter.

Compression Fittings

The big advantage of compression fittings is that, in theory at least, they are easy to undo if you want to make alterations to your plumbing. In practice they can be very stubborn and you may damage the pipework by using too much force in trying to free them.

For a start, make sure you know which way to undo the cap nut; your spanner should be rotating anticlockwise as you look along the pipe towards the fitting. Secondly, make sure the fitting is braced – by another spanner on the fitting's body for straight couplers, or by some other means for tees, elbows and the like. Try tightening the nut fractionally first to free it. If that fails, play heat from a blowlamp or hot air gun lightly over it, and try again. Lastly, try applying some penetrating oil at the point where the cap nut meets the fitting.

Use the same techniques to free seized-up connectors to radiators, again making sure you know which way to turn the connecting nut. Here too, bracing is important; too much force can rotate the radiator valve and fracture the circuit pipework connected to it, necessitating a major repair job.

Tap Connectors

There are two common problems in removing tap connectors (and also the back-nuts that secure the tap body itself to the appliance it is serving) even when you use the right tool – a crowsfoot or basin spanner.

The first is that space is often very tight, preventing you from being able to exert any leverage; it's often difficult even to see what you are doing. The second is that leaks and drips have often caused a build-up of scale and corrosion on the tap tails which can lock the threads of the fittings on to the tap tail and make them very difficult to undo.

As far as the space problem is concerned, try to improve access as much as you can. For example, there may be enough play on the pipework for you to undo any wall or floor fixings and to ease the appliance away from the wall slightly. If this is not possible, locate your crowsfoot spanner on the nut, wedge it in place with a length of timber placed between the bottom of the spanner and the floor, and use an adjustable spanner or wrench to grip and turn the spanner handle.

If this fails, turn to penetrating oil again, trying to trickle it over the nuts so it can penetrate the locked threads. Lastly, try heat to free the fixings, but be careful with blowlamps and hot air guns which can crack ceramic basins and melt plastic baths if used carelessly. Shield the appliance as best you can with glass-fibre matting, then play heat gently over the fitting and allow it to cool down before trying the spanner on it again.

As a last resort, saw through the supply pipework, disconnect the waste pipe and remove the appliance so you can tackle the nuts more readily.

Tap Covers and Handles

When a tap needs rewashering, it is often very difficult to remove the tap handle (or the tap cover on older shrouded-head taps). Many taps have hidden grub screws under plastic indicator discs; check this by prising the discs up before trying other methods. Next, try tapping slim wooden wedges between the handle and the tap base. If this fails, you may have to saw through the tap handle to free it, and fit a new one to replace it.

With shrouded taps, try pouring boiling water over the shroud or trickling penetrating oil down the tap spindle, then grip it with an adjustable wrench and try to unscrew it. Protect the plating with cloth pads.

Lastly, if the shroud can be undone but the handle will not come off, try using the shroud itself to lift it. Open the tap handle fully, undo the shroud and place offcuts of wood under each side to force the shroud up against the handle. Then close the tap; this should force the handle up off its spindle.

INDOOR JOBS

Most of the major plumbing projects you are likely to carry out are indoor ones, concentrated in the main water-using areas of the house – the kitchen, the bathroom and the WC.

In this chapter you will find details of how to carry out thirteen different plumbing projects, ranging in complexity from fitting new taps to installing showers and plumbing in washing machines to re-placing a bathroom suite. In each case you will find background information on the job, a check-list of what you need in order to carry it out, a summary of what the job involves and detailed diagrams to help you make the connections.

What to Tackle First

Plumbing jobs fall into two broad cate-gories: things you need to do, and things you want to do. In the former category come jobs like replacing old taps, while the latter group includes tasks such as fitting an extra wash-basin in a bedroom or installing a waste-disposal unit.

Whatever you plan to do, the best way of drawing up a plan of action is to list the facilities you want in each room of the house in turn.

Kitchen Fit new taps to the existing sink, perhaps replacing old pillar taps with a new mixer tap and using the spare tap hole to fit a sink rinsing spray.

Fit a new sink, taking the opportunity to have an extra washing-up bowl or a different drainer configuration, and per-haps installing a waste-disposal unit underneath the second sink.

Provide new plumbing and waste water connections for a washing machine and/ or a dishwasher, perhaps as part of a larger-scale kitchen remodelling.

Utility room If you are lucky enough to have one, it may be preferable to provide plumbing arrangements here for the wash-ing machine and use the floor space it used to occupy in the kitchen for a dish-washer.

If the utility room has a seldom-used sink, consider removing it and installing a shower cubicle instead – invaluable for soothing hot, tired gardeners and for hos-ing down dirty children.

Bathroom Here the sky really is the limit. Changes can range from simple tasks like fitting new taps to wholesale replace-ment (and rearrangement, if this is necess-ary or feasible) of the main items of bath-room equipment. For example, fitting a corner bath might create enough space to add a bidet or include a shower cubicle. Replacing an existing pedestal basin with a vanity unit could provide useful extra storage space.

Separate WCs and cloakrooms Fit a new low-level WC to replace an old-fashioned type. A slim-line cistern could save valuable space, perhaps allowing the installation of a small recessed wash-basin. Upstairs, a new double-siphonic pan could reduce plumbing noise. You could even partition off part of a hall, landing or bedroom and install an extra WC (see page 47).

Bedrooms If existing plumbing per-mits, consider fitting a wash-basin or even a shower cubicle in one or more bedrooms. Such an installation could have a dramatic effect on the early-morning traffic jams if your home has only one bathroom.

In the loft It is worth bearing in mind that you may need to consider extra water storage facilities to cope with the demand from any extra appliances you install.

Fig 69 (*above*) Getting water to where you want it is rarely a problem, since supply pipework is relatively unobtrusive.

Fitting New Taps

Fitting new taps not only gives an old bath, basin or sink a new lease of life, spelling the end of hard-to-turn handles and dripping spouts; it is also an integral step in installing a new bath, basin or sink (*see* also pages 38 to 45 and 56 to 58).

Your first decision is what type of tap to choose, for function as well as style. You can have single taps, two-hole mixers (so-called because they fit over a pair of mounting holes) or mono-bloc mixers, which have two small-diameter inlet pipes designed to pass through a single mounting hole. Mixer taps may also incorporate a concealed pop-up waste mechanism linked to the appliance's plug.

What to do

If you are replacing existing taps, your biggest problem will be in disconnecting the water supplies to them and in removing the backnuts holding the taps in place. This is because you generally have restricted access to them and the nuts may be hard to undo (*see* page 34 for tips on loosening them).

Start by turning off the water supply (*see* pages 32 and 33) and open the taps. Then use your crowsfoot spanner or Taptool to undo the connectors linking the supply pipes to the tap tails. Have some absorbent cloths handy to catch the inevitable drips. Use the same tool to loosen the

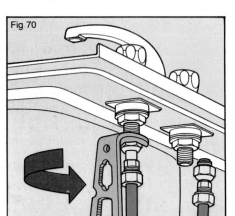

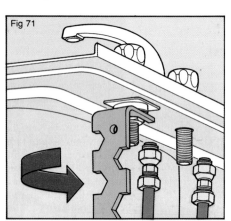

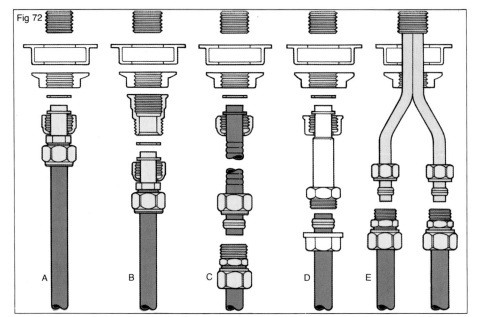

Fig 70 Undo the old tap connectors.

Fig 71 Unscrew the back nuts holding the taps in place, and lift them out.

Fig 72 You may be able to reconnect the supply directly (A), but short tap tails will need a shank adaptor (B) or hand-bendable copper pipe (C). Use plastic connectors (D) on plastic tap tails, and reducing couplers (E) on mono-bloc mixers.

Fitting New Taps

backnuts; brace single taps with scrap wood to stop them turning as you do this. Lift the old tap out and remove any traces of old sealing compound from round the mounting hole.

Now compare the tails of your new taps with the old ones. If they are shorter, you may need a shank adaptor to bridge the gap. Alternatively you can use a length of hand-bendable pipe to make the connection. If they are plastic, you may have to use a plastic tap connector to avoid damaging the tap threads. And with mono-bloc mixers, you will need reducing connectors to link the narrow 10mm diameter tap tails to the supply pipework.

Once you have sorted out your requirements, you can fit the new taps. Start by positioning the tap in the mounting hole, with washers or gaskets in place on the tap tail. Secure it with the backnut (and a top-hat washer on thin surfaces) and then reconnect the supply pipework as dictated by the pipe types and layout. Restore the water supply and check for any leaks from the connections.

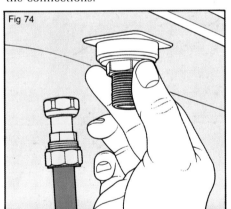

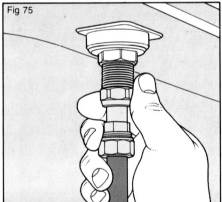

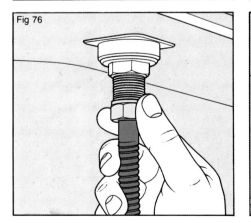

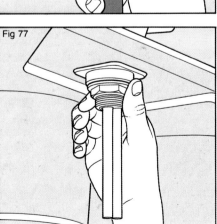

Fig 73 (*above*) Your choice lies between single taps (squat for baths and basins, taller for sinks), two-hole mixer taps or mono-bloc mixers. With two-hole mixers the spacings of the tails are different for baths, basins and sinks.

Fig 74 Pass the tap tail through the mounting hole and tighten up the backnut.

Fig 75 If possible, reconnect the supply pipe directly. Replace the connector washer if it is damaged.

Fig 76 Otherwise use a shank adaptor or hand-bendable copper pipe.

Fig 77 With mono-bloc mixers, both tails pass through one hole.

37

Fitting a Pedestal Basin

Pedestal basins come in two parts – the basin itself, and a matching ceramic pillar which supports the basin and hides the plumbing. The basin is secured to the wall surface via screws driven through holes in the underside of the basin moulding, and the pedestal is then positioned beneath it and screwed to the floor so everything is rigid once installed.

When choosing a basin, decide whether you want pillar taps or a mixer tap (usually complete with a pop-up waste), since basins are made with one or two tap holes to suit the tap type. You will also need to buy tap connectors or reducing couplers, a waste outlet and a trap.

What to do

If the new basin is replacing an existing one in the same position, your first job will be to remove the old basin. Turn off the water supply (see page 32) and disconnect the old supply pipes. It's generally quicker and easier to saw through the supply pipes beneath the old basin, rather than wrestling with the old tap connectors. Cap the pipes with compression stop-ends or fit mini-service valves (see Tip). Disconnect the old trap, undo the old fixing screws and lift the basin away. Try penetrating oil or brute force if the screws are stuck.

What you need:
- basin and pedestal
- pillar taps plus plug and chain or
- mono-bloc tap with pop-up waste
- waste outlet plus washers
- 15mm flexible pipes with ½in tap connectors
- reducing couplings for 10mm tap tails
- 32mm trap with universal connector
- plumbing tools

CHECK
- that with some Continental pop-up waste mechanisms you have an appropriate adaptor to connect it to the trap

TIP
Fit mini-service valves (gate valves) to the cut ends of the existing supply pipes when you remove the old basin. This not only allows you to restore the water supply to the rest of the bathroom while you finish the job; it also means you can re-washer the taps in future without having to turn the water off at the mains

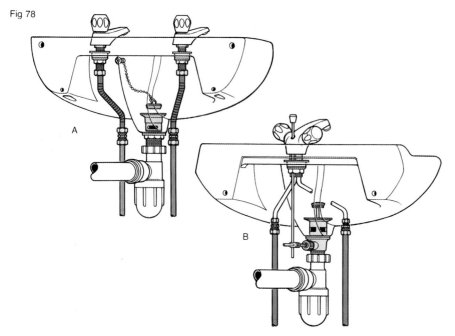

Fig 78

Fig 79

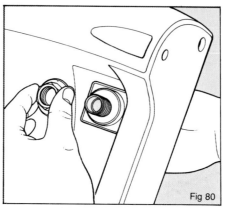

Fig 80

Fig 78 Choose pillar taps and a conventional plug (A), or a mono-bloc mixer with pop-up waste (B).

Fig 79 Set the basin on its pedestal so you can see how the pipes will run.

Fig 80 Then fit the taps, waste, plug or pop-up waste mechanism to the basin.

Fitting a Pedestal Basin

Set the new pedestal and basin in place so you can see what adaptations will be needed to the supply and waste pipes. Then fit the tap(s), waste outlet, pop-up waste (if fitted) or plug and chain to the new basin, and add the trap to the waste outlet. If you are using flexible connectors to link the tap tails to the supply pipework, screw one on to each tap. With mono-bloc taps, fit reducing couplings to the tap tails and connect flexible pipes to them if they won't reach the old pipes.

Replace the basin on its pedestal and check that it is level. Mark the positions of the fixing holes, drill them and push in the wall-plugs. Next, get a helper to hold the basin and pedestal away from the wall slightly while you make the plumbing connections, so you have more room in which to work. When everything is connected up, ease the basin back into position and drive the screws through the basin into their wall-plugs. Finally, secure the pedestal.

Fig 81 (*above*) Pedestal basins have clean, elegant lines and occupy very little floor space.

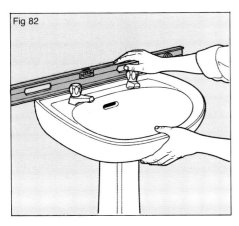

Fig 82

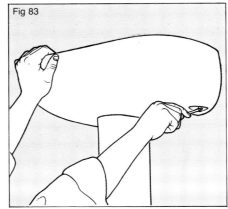

Fig 83

Fig 82   Set the basin back on its pedestal and check that it is level.

Fig 83   Mark the fixing holes on the wall, then drill and plug them.

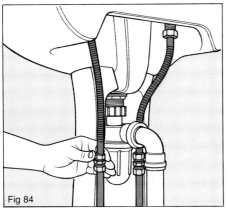

Fig 84

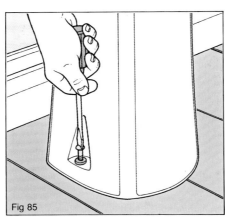

Fig 85

Fig 84   Ease the basin and pedestal away from the wall slightly, and make the supply and waste connections.

Fig 85   Ease the basin back into place, secure it to the wall and screw the pedestal to the floor.

Fitting a Wall-Hung Basin

Wall-hung basins are a good choice where space is tight and you want to keep the floor area uncluttered. Most modern types are supported on a wall-mounted bracket which fits round the waste outlet; some also rely on fixing screws, or slot over wall-mounted anchor bolts (see Fig 86).

What to do

Remove the existing basin (see page 38), and make up the new basin with taps, waste outlet and so on as for a pedestal basin. If a template is supplied, use it to mark the fixing positions; otherwise get a helper to hold the basin while you mark them directly. Next, secure the bracket to the wall (see Check), and place the basin on it. Depending on the bracket type, secure it with screws, locate it over projecting bolt heads or engage adjustable hooks (see Fig 86). Complete the job by connecting the supply pipework to the tap tails and linking the existing waste pipe to the new trap.

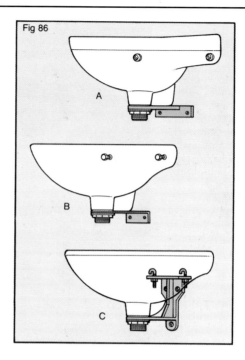

Fig 86

A

B

C

What you need:
- basin and bracket
- pillar taps plus plug and chain *or*
- mono-bloc tap with pop-up waste
- waste outlet plus washers
- 15mm flexible pipes with tap connectors
- reducing couplings for 10mm tap tails
- 32mm trap with universal connector
- wall fixings
- plumbing tools

CHECK
- that the wall is strong enough to support the basin. Use No 12 fixing screws and wall-plugs on brick- and blockwork; on a stud wall either screw to the studs or fit a support batten between them

TIP
Set the front of the basin about 825mm (32½in) above floor level for adults, lower for children.

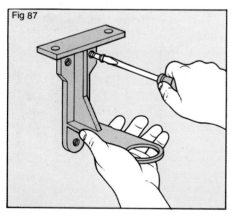

Fig 87

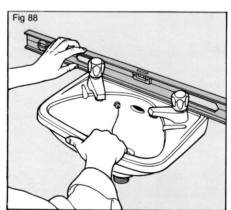

Fig 88

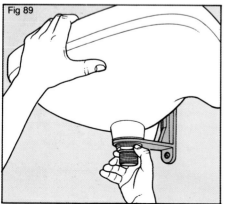

Fig 89

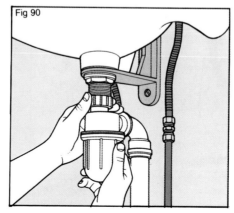

Fig 90

Fig 86   Wall-hung basins are supported on a sturdy bracket, and may be held against the wall by screws (A), projecting bolt heads (B) or adjustable hooks on the wall support bracket (C).

Fig 87   Fit the bracket securely to the wall.

Fig 88   Set the basin on it and check that it is level.

Fig 89   Secure the basin with its backnut.

Fig 90   Connect the supply and waste pipework.

Fitting an Inset Basin

Inset basins can be installed into the work surface of a vanitory unit, or can be built into a work-top with a tiled or other surface. Unlike other basin types, you have a choice of acrylic or ceramic basins; the former are cheaper.

What to do

Start by preparing the work-top or other surface into which the basin will fit by cutting a hole in it with a jigsaw or padsaw. A template is usually provided with the basin. Double check the measurements before starting to cut the hole. Then test the fit of the basin.

Attach the taps, waste and so on to the basin and place mastic or sealing strip round the perimeter of the hole. Lower the basin carefully into place and press it down firmly all round to ensure a waterproof seal. Check that it sits squarely, tighten the fixing clips underneath to hold it securely in position and connect up the pipework.

Fig 91 (*above*) Inset basins can be fitted to your own work-top, or to self-assembly vanitory units.

> **What you need:**
> - as for wall-hung basin (*see* page 40) *plus*
> - mastic or sealing strip
> - fixing clips

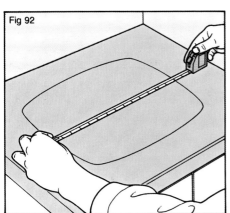

Fig 92

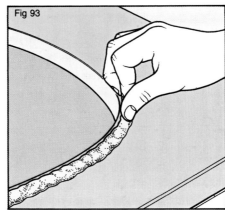

Fig 93

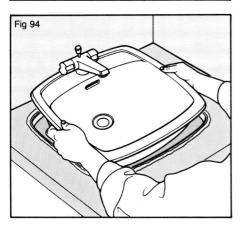

Fig 94

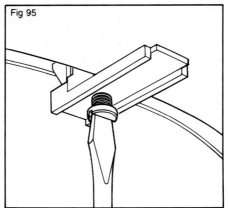

Fig 95

Fig 92   Use the template to mark the hole position on the work-top. Double-check the measurements.

Fig 93   Cut the hole and place mastic or sealing strip round it.

Fig 94   Set the basin squarely in the hole and press down all round.

Fig 95   Secure the basin to the underside of the work-top with clips. Then connect the pipework.

Fitting a Bath

Although the traditional rectangular bath is the most popular type, corner baths of various shapes are increasingly popular because they can allow a more practical use of floor space, especially in small bathrooms. They are no more difficult to install than other appliances; the biggest problem is usually getting them into the bathroom in the first place.

Whirlpool baths are also becoming more widespread, but installing one is best left to a qualified plumber because of the need to ensure that the wiring to the pump is electrically safe. They also need soft (or softened) water to prevent a build-up of scale from blocking the nozzles.

What to do

If the bath is a replacement for an existing one in the same position, your first job will be to remove the old bath. Turn off the water supply (see page 32), remove the bath panels and disconnect the old supply pipes and the waste. As with basins, it is usually quicker and easier to saw through the old supply pipes, and to cap them off with mini-service valves (see Tip on page 38). Undo any screws holding the bath to the floor. You should be able to lift an old pressed steel or plastic bath out in one piece, but it's often simplest to break up an old cast iron bath unless you plan to

(see page 32), (see Tip on page 38).

What you need:
- new bath with legs (metal types) or cradle (plastic types)
- bath panels
- new taps oblique mixer or shower mixer
- waste/overflow kit – pop-up or plug/chain
- 22mm flexible pipes with ¾in tap connectors
- 40mm trap with universal connector
- fixing screws
- plumbing tools

POP-UP WASTES
Fit the control knob into the overflow hole and the waste mechanism to the outlet hole. (See Fig 97.)

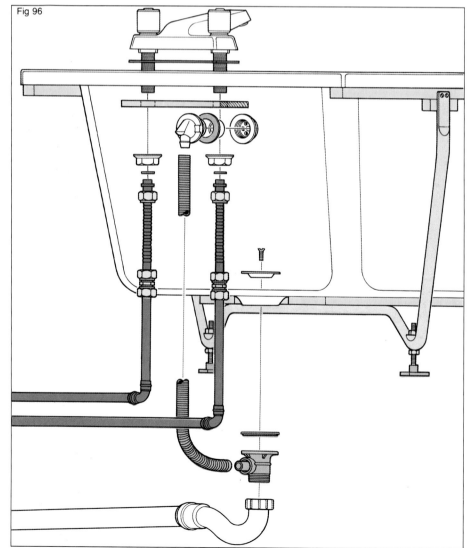

Fig 96

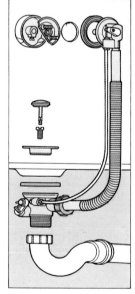

Fig 97

Fig 96 Typical plumbing and waste connections for a metal or plastic bath with mixer taps. The overflow is linked to the waste outlet by a flexible hose.

sell it. Cover the bath with a blanket and smash it into manageable pieces with a sledgehammer. Wear gloves and goggles, and take care – sharp slivers of enamel may fly off, and the edges of broken pieces can be extremely sharp.

Re-route the supply and waste pipes at this stage of the job if you plan to install the new bath in a different position in the bathroom.

With plastic baths, start the installation by assembling the cradle. With steel baths, attach the feet with the adhesive pads provided. Connect up the waste and over-flow fittings, sealing the holes with washers or mastic. Tighten centre screws with a coin and check the operation of pop-up wastes. Then fit the taps (use a reinforcing plate on acrylic baths) and add flexible pipe to each tail so you can connect the supply pipes easily.

Position the bath and check that the trap will fit. Adjust the feet so the bath is level, and extend supply and waste pipes if necessary before securing the wall and floor brackets. Check all connections.

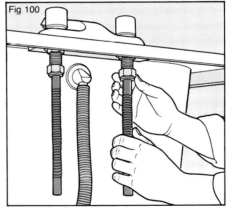

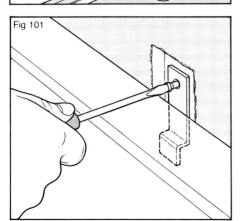

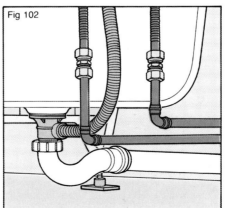

Fig 98 (*above*) Modern baths usually come complete with matching side and end panels, but you can box them in yourself if you prefer.

Fig 99 Turn a new plastic bath upside down and assemble its cradle.

Fig 100 Fit the taps, waste outlet and overflow hose, and add flexible pipes to connect the tap tails to the existing supply pipes.

Fig 101 If wall brackets are required, chip away the plaster and screw them to plugged holes.

Fig 102 Connect the overflow to the waste outlet and attach the trap. Connect in the waste pipe and link the flexible pipes to the supply pipework.

Fitting a Bidet

Installing a bidet is exactly like fitting a wash-basin – at the bidet end of things at least. There are two types of bidet: one has an over-rim supply, usually via a three-hole mixer tap with pop-up waste; the other has a special bidet mixer tap with a diverter valve which fills the bidet through outlets in the rim or via an ascending spray. The over-rim type can be supplied with hot and cold pipes teed off the existing bathroom services, but the rising spray type must have its own independent supplies run from the hot cylinder and the cold water storage tank to avoid any risk of back-siphonage. This is a requirement of the water supply by-laws.

What to do

Start by fitting the waste outlet and trap to the bidet, then attach the mixer tap and connect up the pop-up waste mechanism. Next, offer up the bidet to its chosen position so you can mark the lines of the supply and waste pipe runs.

With an over-rim supply bidet, tee the supply pipes off the existing bathroom hot and cold pipes at the most convenient point. For ascending-spray bidets, run a new cold feed from a tank connector at the cold water storage tank, and tee off a new hot supply pipe immediately above the top of the hot cylinder.

What you need:

- over-rim supply bidet
- single taps, waste outlet, plug/chain *or*
- mixer tap with pop-up waste
- ascending-spray bidet
- spray mixer tap with diverter valve
- flexible pipe with tap connectors
- fixing screws
- 32mm trap with universal connector
- plumbing tools

TIP
Keep the pipe for an over-rim supply bidet short by teeing into the nearest point on the existing hot and cold pipework. (*See* Fig 103.)

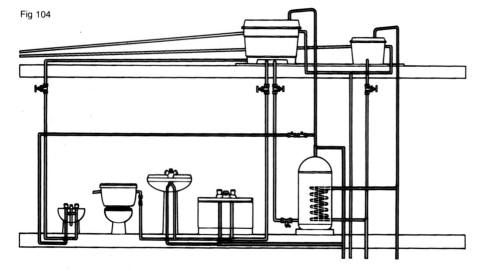

Fig 104

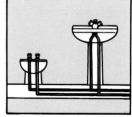

Fig 103.

Fig 104   Separate hot and cold supply pipes must be run to an ascending-spray bidet.

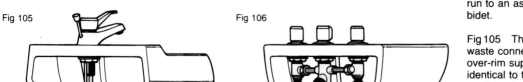

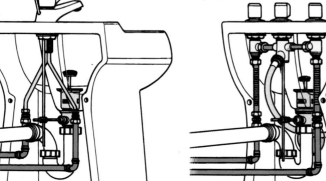

Fig 105

Fig 106

Fig 105   The plumbing and waste connections to an over-rim supply bidet are identical to those for a wash-basin. You will need reducing couplings to connect up the tails of mono-bloc mixer taps.

Fig 106   For an ascending-spray bidet, the supply and waste connections are the same as in Fig 105, with a flexible hose supplying the spray.

Fitting a WC

If you decide to replace an existing WC, you must first decide what type to install. Most new WCs these days have close-coupled cisterns, with the cistern sitting on the back of the WC pan (see Fig 109). Your existing WC, however, may well have a separate cistern and pan, with the two connected by a short length of flush pipe (see Fig 108). You also have to think about the pan design. The cheapest suites have what is known as simple wash-down action, while more expensive types have siphonic action which is generally quieter and also more efficient that the basic wash-down type. (see Pan Types on page 47 for more details).

Variations on the WC theme which you might like to consider as alternatives are suites with slimline cisterns, ideal for small bathrooms; wall-hung WCs (and matching bidets) which leave the floor surface clear of pedestals; and boxed-in cisterns, which allow you to conceal supply and soil pipes neatly. All are available with flush mechanisms that reduce the volume of water used to the empty pan.

Fig 108

Fig 109

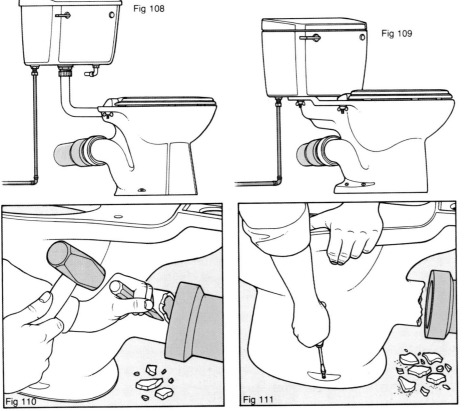

Fig 110

Fig 111

Fig 107 (*above*) Modern WC suites are low, compact and attractively styled to match other bathroom appliances, especially bidets which often sit alongside them.

Fig 108 Old-fashioned WCs often have a separate cistern connected to the pan by a flush pipe.

Fig 109 New WC suites are usually the close-coupled type, with the cistern sitting on the rear of the pan.

Fig 110 Use a cold chisel to break the outlet of the old pan and free it from the soil pipe.

Fig 111 Then unscrew the pan fixings and lift it aside. Clean out the soil pipe socket thoroughly.

45

Fitting a WC

What to do

Once you have chosen your new suite, your first step must be to remove the existing WC (unless you are creating a bathroom from scratch).

Start by turning off the cold supply to the old cistern, flush it and mop out any remaining water. Since you will probably have to re-position the supply in order to connect it to the new cistern, simply cut through the supply pipe and fit a cap nut or a mini-service valve so you can restore the water supply to the rest of the bathroom. Cut the overflow too. Undo any fixings holding the cistern to the wall. Unscrew the flush pipe at the base of the cistern, or undo the nuts holding it to the pan if it is the close-coupled type, and lift the cistern aside.

Now you can tackle the old pan. Empty the trap as much as possible with a plunger (or by plunging vigorously with a lavatory brush), then stuff in some rags to absorb the water that is left. If the pan outlet is cemented into the soil pipe, either chip away the mortar or carefully crack the pan outlet; it doesn't matter if you break this, whereas cracking the soil pipe could mean an awkward repair job. If a plastic connector was used, it will simply pull off once you have freed the pan.

Free the pan next. If it is screwed to the floor, you may need to apply penetrating oil to free the old screws. If it is bedded in mortar, drive a cold chisel beneath pan and floor at several points to prise the pan up. Lift the pan aside and carefully chip any mortar or remains of the old pan outlet out of the soil pipe socket. Finally, tie a plastic bag over it to keep drain smells out.

Assuming that the new pan will sit in the same position as the old one, bring it in and set it in place so you can see what type of pan connector you will need to connect it to the soil pipe, and whether any extensions will be required.

Start by assembling the siphon unit, the ball valve and the lever assembly to the cistern, following the instructions included with the fittings pack. Make sure you

Fig 112

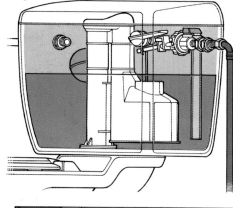

Fig 113

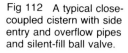

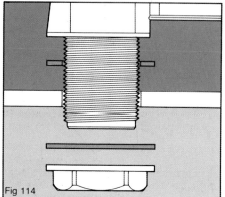

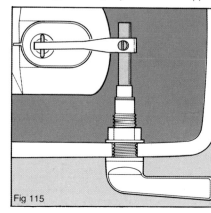

Fig 112   A typical close-coupled cistern with side entry and overflow pipes and silent-fill ball valve.

Fig 113   The internal workings of a typical slim-line cistern. Note the bottom-entry supply and overflow pipe.

Fig 114   Check that both internal and external washers are fitted before securing the siphon unit to the cistern with its backnut.

Fig 115   Fit the lever and link it to the siphon mechanism.

fit the internal and external washers to the siphon unit before securing it with its backnut. On ceramic cisterns, small china blanking plugs are provided to fill any unused holes. On plastic cisterns you generally have to cut the holes you require using a hole saw.

You can now push the WC connector on to the soil pipe and offer the pan up to it. If it fits properly, screw the pan to the floor. Check that it is level, packing underneath it if necessary.

Next, offer up the cistern so you can mark the positions of the wall fixings. Drill and plug the holes and reposition the cistern. Tighten the wing nuts holding it to the pan, then secure it to the wall.

Extend the supply pipework to the new cistern and connect it to the tail of the ball valve with a tap connector. Finally, make up the new overflow pipe, connecting it to the cistern and running it through the wall. It is generally easier to remove the old pipe completely than to try to joint new to old. Double check all your connections, then restore the water supply to fill the cistern.

If you intend to reposition the WC somewhere else in the bathroom, you can do so by running new soil pipe round the room from the original WC position. However, remember that the pipe run must not exceed 6m (19ft 6in) in length, and should have a minimum slope of 9mm per metre downwards towards the stack.

PAN TYPES

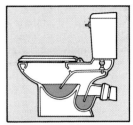

Fig 121   Double-trap siphonic.

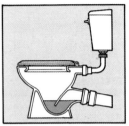

Fig 122   Simple wash-down.

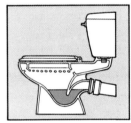

Fig 123   Single-trap siphonic.

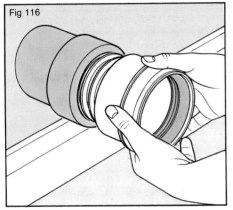

Fig 116

Fig 117

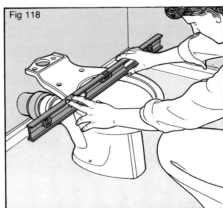

Fig 118

Fig 119

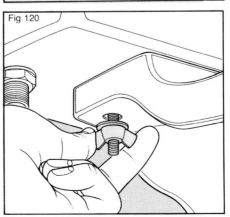

Fig 120

Fig 116   Remove the plastic bag from the soil pipe and fit the new pan connector.

Fig 117   Offer up the pan so its outlet engages fully in the connector.

Fig 118   Check the pan is standing level – insert packing such as old tiles beneath it if necessary. Then screw it to the floor.

Fig 119   Extend the supply pipework if necessary.

Fig 120   Fit the cistern and secure it to the pan with wing nuts. Finally, connect the supply and overflow pipes to complete the job.

Fitting a Shower Tray

Unless you plan to install a shower over the bath, you will need a shower tray and some sort of shower enclosure. You can create the enclosure yourself, or buy a self-assembly shower cubicle (see pages 50 and 51). In the latter case it is sensible to choose tray and cubicle together so you can be sure the one will fit the other, but this is less important if you are planning a home-made enclosure, since you can tailor-make this to fit any size of tray.

Shower trays are available in ceramic or plastic versions; the latter are either acrylic (cheaper) or glass-reinforced plastic (GRP – more expensive). The simplest trays measure between 750mm (30in) and 900mm (35½in) square, but you can also buy rectangular trays and rounded quadrant shapes. Plastic trays are light in weight and usually have adjustable feet, making installation and levelling easy, while ceramic trays are heavier but feel sturdier underfoot. Most are between 200mm (8in) and 300mm (12in) high, and have a 40mm (1½in) diameter waste outlet to which the trap and waste pipe are connected.

What to do

Start by working out how the waste pipe from the shower is going to be run, since this will affect the level at which you install the shower tray. The pipe needs a slight fall to ensure that it drains properly – the longer the pipe run, the shallower the slope, up to a maximum of 3m (10ft). For pipe runs over 3m (10ft) you will have to add an anti-siphon trap to stop the trap emptying as the waste pipe runs dry. The waste pipe can be taken to a hopper or gully outside, or connected to a soil stack (see page 29).

You will now know whether the tray can be sited directly on the floor surface or whether it will need to be raised slightly on a plinth – this is the most likely situation. If you need a platform, build it on sturdy timber for plastic trays; you can use bricks for ceramic ones and box them in. Design it to leave a removable access panel so you can reach the trap if it becomes blocked with hair (a common problem with showers), and include a step in

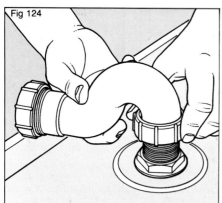

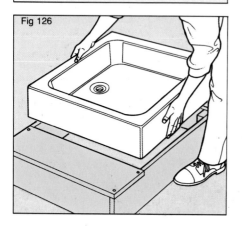

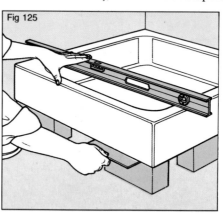

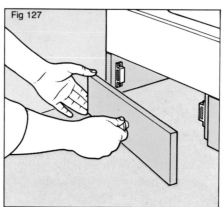

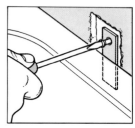

Fig 128.

Fig 124   Fit the waste outlet, using sealant to ensure a watertight joint, and then fit the trap.

Fig 125   If you need a platform, you can use bricks with a ceramic tray. Add packing to get the tray level.

Fig 126   With plastic trays, build a timber platform and incorporate a step up to the tray if you wish.

Fig 127   Leave an access hatch so you can get at the trap if it becomes blocked.

Fitting a Shower Tray

the framework if you wish to make access to the tray easier. Make sure the framework is level; you will not be using the tray's adjustable feet in this situation.

You can now install the tray. Start by fitting the waste outlet and then attach the trap. With a floor-level installation, you may have to cut away a section of the floor to accommodate the trap – do this next. Set the tray in position and use the adjustable feet (if fitted) and get it absolutely level. This type of tray usually has clip-on side panels. Other one-piece plastic trays generally incorporate anchor straps which must be screwed to the floor, the supporting platform or the enclosure walls. No fixings are needed with ceramic trays.

Once the tray is in place, all that remains is to connect up the waste pipe run in preparation for the cubicle to be fitted or the enclosure completed.

If the waste pipe run is long, it is a good idea to include a rodding eye near one end of the run so that any blockages which occur in the pipe can be cleared without having to dismantle it.

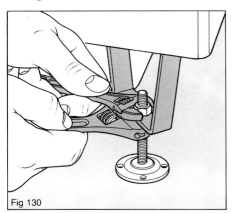

Fig 130

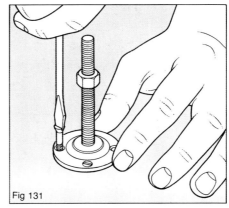

Fig 131

Fig 129 (*above*) A shower above a tray is both space-saving and economical.

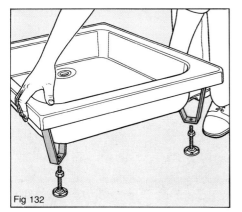

Fig 132

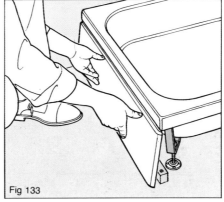

Fig 133

Fig 130   On trays with adjustable feet, use two spanners to adjust the height and lock nuts.

Fig 131   When you have the tray level, mark the fixing holes and lift the tray off its feet so you can secure them to the floor.

Fig 132   Lift the tray back into place and fit the lock nuts.

Fig 133   Finally, clip the tray panels into place. As an alternative, use plywood panels covered with tiles or the bathroom floor-covering.

49

Fitting a Shower Cubicle

Self-assembly shower cubicles come in a wide range of types, styles and finishes. The commonest consists of a series of aluminium extrusions to form the cubicle frame, and sheets of textured (and sometimes coloured) acrylic or other plastic to form the walls and door. More expensive types use glass, and it is essential to check that this is safety glass – toughened or ideally laminated – so that a fall in the shower cannot cause a serious accident. There are also all-in-one cubicles with solid wall panels, designed to form a watertight enclosure without the need for do-it-yourself waterproofing, but they are comparatively expensive and can be awkward to manoeuvre in confined spaces.

Square enclosures may have one, two or three sides; choose whichever suits your installation. There are also curved cubicles, designed to fit over quadrant-shaped trays. Doors may slide, fold or hinge outwards; if possible check that the door action allows easy access into and out of the enclosure, and that the door moves and closes easily.

What to do

Start by unpacking your shower kit and read the instructions so you can identify all the parts and check that nothing is missing. The assembly sequence obviously varies from brand to brand, but most installations follow a similar sequence of operations. Details on how to fit a typical two-panel corner cubicle are given below.

Step one is to measure the size of the shower tray, in case you have to make any adjustments to the kit components to get a good fit. Then mark the positions of the wall uprights on the walls, checking that they are truly vertical – a spare pair of hands is helpful at this stage. Drill and plug the fixing holes and secure the uprights in place. If the walls are out of true, use packing pieces or the profiles included in the kit to compensate for this.

Offer up one of the shower wall panels (which might include a door) and check that it fits the tray. If it doesn't, you will have to cut down the glazed panel and the top and bottom frame members to size;

What you need:

- cubicle kit
- wall fixings
- silicone mastic
- general-purpose tools

CHECK

- that the cubicle you choose is big enough to fit your tray
- that all components are present when you unpack the kit
- that your walls are square. If they are not, pick a kit with adjustable profiles

Fig 134

Fig 135

Fig 136

Fig 137

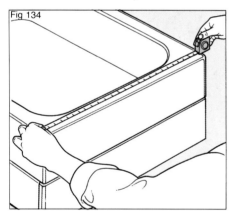

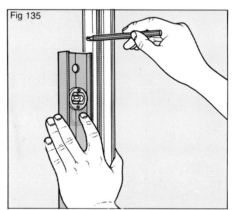

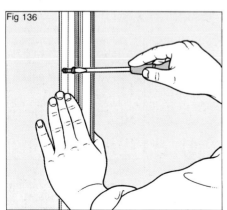

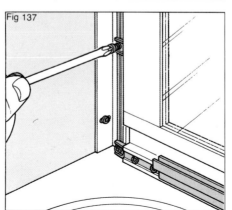

Fig 134 Start by measuring the tray to see if any adjustments are needed to the kit panels.

Fig 135 Offer up the wall uprights, check that they are vertical and mark the fixing screw positions.

Fig 136 Screw the uprights to the walls.

Fig 137 Offer up the side panels to the wall uprights and after any necessary trimming screw them in place.

Fig 138 Connect the panels at top and bottom with the corner blocks.

Fig 139 Hang the door if it is separate, and adjust it so it operates smoothly.

the instructions will tell you how to do this. When you are happy with the fit, screw the panel to the wall upright.

Repeat this operation for the other side panel. The two are usually linked by special corner joints at the top and bottom. Fit these joints next and check that the whole assembly is standing perfectly square on the tray – if it isn't, the door(s) will not operate properly. Finish off by locating the door on its track or brackets if it is a separate component, and make any necessary adjustments to ensure that it operates smoothly.

Complete the job by sealing all the joints between the tray and the cubicle, and between the cubicle sides and the room's walls using silicone mastic.

If the back walls of the enclosure are to be tiled after the cubicle is installed, take great care over the job to ensure that leaks cannot occur. On plasterboard walls, seal the board surface with solvent-based paint first to prevent water penetration. Use a generous layer of waterproof tile adhesive and bed the tiles firmly into it. Grout the

joint lines, either with waterproof grout or, on stud walls, with the same silicone mastic that you used to seal the cubicle joints. This stays flexible and so can cope with any movement in the wall structure which might crack rigid grout and encourage leaks.

The assembly sequence is broadly the same for curved, front-panel and three-sided enclosures.

TIP
On tiled walls, stick masking tape to the tile surface to stop the drill bit from skating.

If you are tiling walls after fitting the cubicle, lay a blanket in the tray so dropped tiles cannot damage it.

Fig 140 (*below*) Shower enclosures come in a wide range of styles. This one has two sliding doors for access.

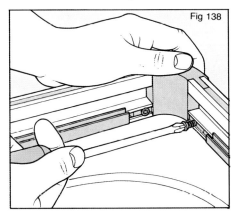

Fig 138

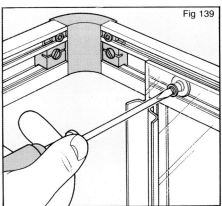

Fig 139

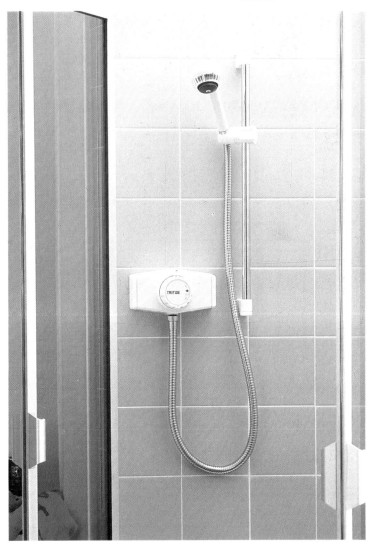

Fitting a Mains Shower

Mains-fed showers are the best choice if you want a powerful flow of water. You can tee the supplies off the existing bathroom hot and cold pipes, but if you do be sure to choose a shower mixer valve that has thermostatic control to maintain the shower temperature if other taps are used while it is running. A manual valve must be fed by independent supplies.

To get a good shower and adequate water flow, you need at least 1m (3ft) of head between the cold tank and the shower rose. Keep pipe runs as short as possible, with slow bends rather than elbow fittings to improve the flow rate. It is a good idea to use 22mm pipe for most of the pipe run.

If the head is inadequate, consider using a pump to boost the flow rate. Fit a twin-impeller type before the mixing valve or a single-impeller type after it, whichever suits your plumbing arrangements best. Automatic switches turn on the power to the pump when the mixing valve is opened. With twin-impeller pumps, independent supplies should be run from the cold tank and hot cylinder. The connection to the cylinder is normally made to the vent pipe, but this can cause air to be drawn down the pipe and into the pump. An alternative is to connect the supply to the side of the cylinder using an Essex flange (see Fig 142).

What to do

Start by checking the available head and working out how long the pipe runs to the shower will be, so you can decide whether you need a pumped supply or not.

If you do not need a pumped supply, run the supply pipes to the mixing valve position, either burying them in the plaster if you have solid walls or hiding them within timber stud partition walls so they emerge to match the mixing valve inlets. Mount the valve and connect up the pipe tails.

If your enclosure has stud partition walls, you can choose a shower mixer valve which is designed to be recessed behind the wall surface – a neater effect than can be achieved with a surface-mounted unit.

What you need:
- mixing valve
- shower head and hose kit
- supply pipework
- connectors to valve
- tank connector for cold supply
- tee or Essex flange for hot supply
- separate pump and flexible connectors
- plumbing tools

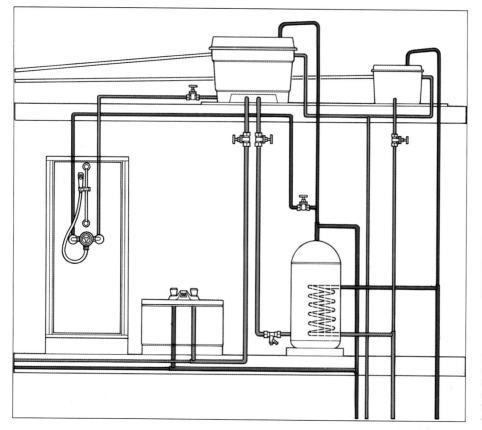

CHECK
- that the head between the cold tank and the shower rose is at least 1m (3ft); if it is not you will need a pump
- that pipe runs are kept as short as possible to maximize water flow rates

TIP
Run the supplies to the shower in polybutylene pipe. This can be bent easily, allowing the run to be made without the need for any joints. This improves the flow and avoids the risk of leaks in buried pipes.

Fig 141 For the best results, run independent hot and cold supplies to the mixing valve from the cold tank and hot cylinder. Fit a gate valve on each pipe.

Fitting a Mains Shower

If you do need a pump, you have a choice of fitting a separate pump (which can usually be concealed under the bath) or of using an all-in-one mixing valve and pump – this is powered by a low-voltage transformer for complete electrical safety. With a separate pump, it is best to leave the wiring and plumbing work to a professional installer.

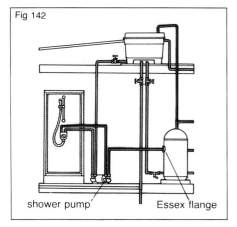

Fig 142

shower pump   Essex flange

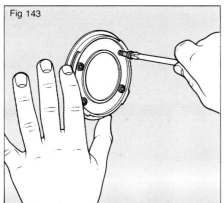

Fig 143

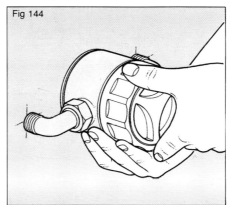

Fig 144

Fig 147 (*above*) The neatest mains shower valves can be flush-mounted to conceal the supply pipework.

Fig 142 With a pumped shower, a connection to the vent pipe may suck air into the pump. Avoid this by connecting the pipe to the side of the hot cylinder with an Essex flange.

Fig 143 Mount the valve base plate on the wall of the shower enclosure.

Fig 144 Offer up the valve body so you can mark the pipe positions.

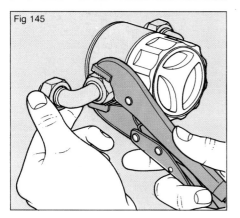

Fig 145

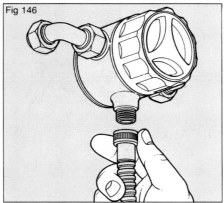

Fig 146

Fig 145 Run in the supply pipework so it emerges through the wall, and connect it to the valve.

Fig 146 Attach the flexible hose supplying the shower head.

Fitting an Electric Shower

An electrically-powered shower is an alternative to a mains shower if there are plumbing complications or if the shower site is remote from the existing supplies. This system takes water at high pressure direct from the rising main and passes it over a heating element and then on to the shower head. The latest models are rated at up to 8.4kW and give quite a good flow rate.

The plumbing to an electric shower could not be simpler, allowing it to be sited in a shower cubicle or over a bath. The electricity supply is a separate radial circuit run from a spare 30-amp or 45-amp fuseway in the main consumer unit.

What to do

On the plumbing front, the simplest thing is to work backwards from the shower position to the point where you will connect into the rising main. As with mains showers, it is a good idea to use flexible polybutylene pipe for buried runs to avoid the risk of leaky joints. Connect the pipework to the shower inlet and run it via a conveniently-sited stoptap to the connection point. Turn off the rising main stoptap, drain the pipe and cut in a connecting tee. If copper pipe is used for the run, it must be cross-bonded to earth via a clamp and earthing cable (see Check).

What you need:
- electric shower
- supply pipework and fittings
- stoptap
- connecting tee
- double-pole switch to match shower rating
- two-core and earth cable to match shower rating (see text)
- plumbing and electrical tools

CHECK
- that metal supply pipes are cross-bonded to earth via a special clamp and an earthing cable (see Fig 150.)
- that a double-check valve is fitted to the supply pipe if the shower rose can be submerged in a bath – to prevent back-siphonage

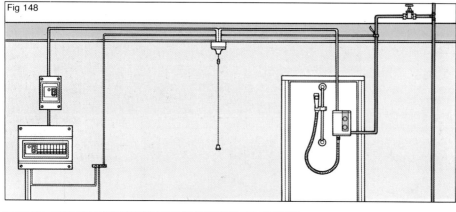

Fig 148

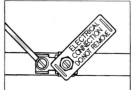

Fig 150

TIP
In hard water areas, fit a water softener on the supply pipe to reduce scale build-up.

Fig 148   Run the water supply to the shower from the rising main via a stoptap. Wire the power circuit from the fusebox to a double-pole switch and on to the shower unit. A residual current device (RCD) on the circuit gives added electrical safety.

Fig 149   Within the shower casing, connect the supply pipe with a straight or swivel tap connector.

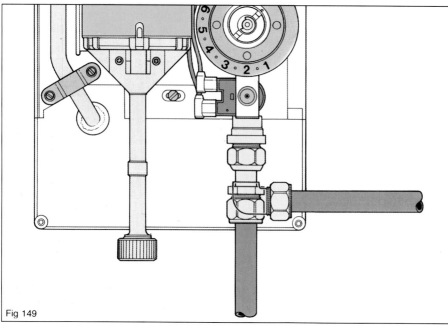

Fig 149

Fitting an Electric Shower

As far as the electrics are concerned, this is a job you may prefer to leave to a qualified electrician, especially if there is no spare fuseway available, unless you are confident of your ability to do the job correctly and safely. A bodged job could kill someone.

The circuit is run in two-core and earth cable. Use 4mm^2 cable for showers rated at up to 7kW and protected by a 30-amp cartridge fuse or miniature circuit breaker (MCB), and 6mm^2 cable for more powerful showers or where rewirable circuit fuses are fitted.

Run the cable from a 30-amp or 45-amp fuseway as appropriate. Take it to the shower via a double-pole switch rated to match the circuit fuse. This must be the cord-operated ceiling-mounted type if it is within 2m (6ft) of the shower or a bath. At the shower itself, run the cable in from behind the unit and connect the cores directly to the heater terminals. Make sure it passes through a waterproofing grommet where it enters the shower casing, so water cannot enter.

Fig 151 (*above*) Electric showers are compact and easy to install in a cubicle or over a bath.

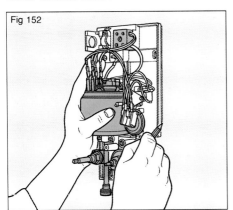
Fig 152

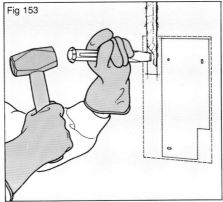
Fig 153

Fig 152   Hold the shower in position so you can mark the fixing positions and plan the cable run.

Fig 153   Chase out the cable run, and clip the circuit cable in place.

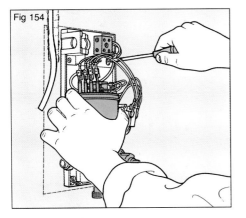
Fig 154

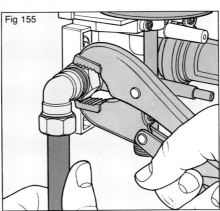
Fig 155

Fig 154   Mount the unit on the wall, using rustproof screws.

Fig 155   Connect the supply pipework and the power supply, fit the casing and test the shower.

Fitting a Sink

Kitchen sinks come in two basic types – lay-on and inset – and in many different shapes and sizes. The lay-on sink sits on top of a standard double- or triple-base unit, while an inset sink is set into a hole cut in a work-top. Both types may have one or two bowls, with one of the bowls often half-sized and ideal for use with a waste-disposal unit (see page 59), and an integral drainer to the left or right of the bowls. Both can be fitted with separate pillar taps or various types of kitchen mixer (see Check).

The water supplies to the sink are conventional, with a branch from the rising main supplying cold water, and hot water coming from the hot cylinder or a water heater. Waste connections can be more complex, depending on how many bowls the sink has and whether connections are required at the trap for a washing machine or dishwasher. If you plan to have a waste-disposal unit, make sure one bowl has a 90mm outlet (see page 59).

What you need:
- new sink
- new taps plus connectors (see pages 36 and 37 for options)
- waste and overflow kit
- silicone mastic
- flexible pipe
- new trap(s) plus waste pipe and connectors
- plumbing tools

CHECK
- that sink mixer taps are the divided-flow type, with mixing occuring as the water leaves the tap
- that metal sinks and supply pipework are cross-bonded to earth. Use the tag on the underside of the sink and a special clamp on the pipework to connect the earth cable, and run it to an electrical earth terminal of a nearby wiring accessory

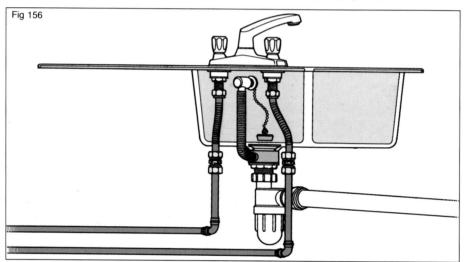

Fig 156

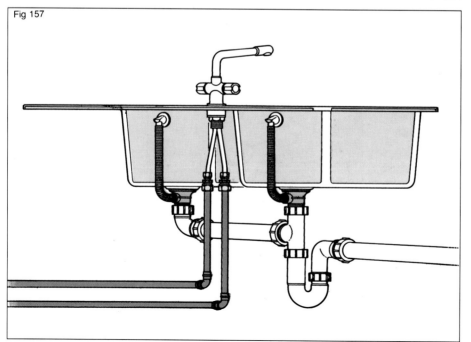

Fig 157

Fig 156 The supply and waste connections beneath a single-bowl sink are simple. Flexible pipe allows the supply pipes to be reconnected easily below bowl level.

Fig 157 With double-bowl sinks, the waste connections can be more complicated. Multi-way traps allow the waste outlets to be linked and fed into a single trap.

56

Fitting a Sink

What to do

Measure up carefully when choosing your sink, especially if you are replacing an old imperial-sized sink: modern metric sinks are slightly different sizes. Also make sure that any obstructions below the sink position will not interfere with the installation.

If you are replacing an existing sink, start by turning off the water supplies. Cut through the supply pipework and disconnect the trap. Undo any screws or clips securing the sink and disconnect the earthing strap linking a metal sink and its supply pipework to earth. Lift the old sink out and set it aside.

Next, fit the new taps, waste and overflow to the new sink. You will need top-hat washers to secure single pillar taps and two-hole mixers. Use the seals and gaskets provided, adding silicone sealant (not plumber's putty) with a banjo-type overflow fitting (see Fig 162). Add flexible pipe to each tap tail so you can make the connections to the supply pipes without having to fiddle around with a crowsfoot spanner when the sink is in position. With mono-bloc taps secured with a large back-nut, use either a large top-hat washer or a steel bracing plate (supplied with the tap) on thin-section steel sinks. Once the tap is in place, the flexible tap tails can be fitted with reducing couplings ready for con-

nection to the supply pipes.

Set a lay-on sink in position and secure it with the retaining clips. Connect the supply pipework, replacing the cross-bonding cable, and fit the new trap(s). Link these to the waste pipe.

With an inset sink, you now have to cut

Fig 158 (*above*) Inset sinks can be installed anywhere in a run of worktop.

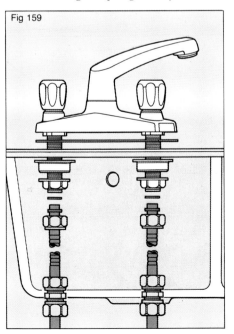

Fig 159

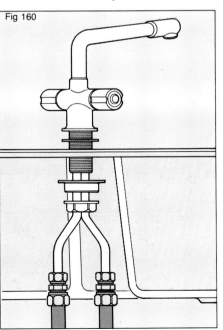

Fig 160

Fig 159 Secure pillar taps and two-hole mixers with a washer or gasket and a backnut, adding a top-hat washer on thin steel sinks. Flexible pipe connects easily to the supply pipework.

Fig 160 Fit mono-bloc taps with a large top-hat washer or bracing plate. Then bend out the flexible tap tails and fit reducing couplings for connection to the old supply pipes.

Fitting a Sink

the hole in the work-top. Use the template provided and cut the hole with a jigsaw. Fit the sealing strip round the perimeter of the hole, set the sink in place and tighten the retaining clips. Then reconnect the supply pipework, fit the trap(s) and complete the waste pipework.

Restore the water supply and check all supply and waste connections for leaks, tightening up any that weep.

All metal sinks and metal supply pipe-work must be cross-bonded to earth for electrical safety. To do this, link the metal tag on the underside of the sink to special earth clamps on both the hot and cold supply pipes, using single-core earth cable. Then connect one of the clamps to the earthing terminal of a nearby electrical wiring accessory, or run the cable back to the house's main earthing point.

TIP
When you cut the old supply pipework, cap off the pipes with mini-service valves so you can isolate the tap for future rewashering or other maintenance.

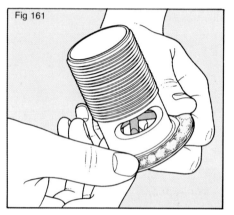

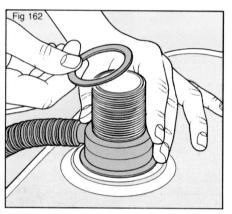

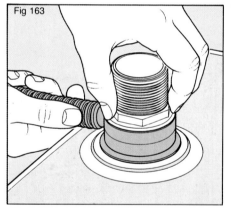

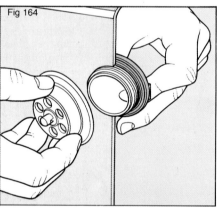

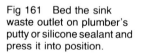

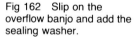

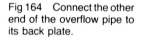

Fig 161 Bed the sink waste outlet on plumber's putty or silicone sealant and press it into position.

Fig 162 Slip on the overflow banjo and add the sealing washer.

Fig 163 Add the backnut and tighten it up fully.

Fig 164 Connect the other end of the overflow pipe to its back plate.

Fig 165 Mount the taps, using gaskets above the sink and top-hat washers below on thin steel sinks.

Fig 166 Set the sink in place ready for connection to the supply and waste pipework. Bed inset sinks on flexible sealant. Secure the sink with screws or clips.

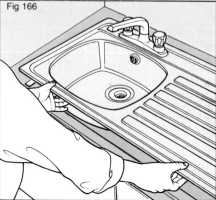

Fitting a Waste-Disposal Unit

Waste-disposal units are an increasingly popular feature today, because of their ability to dispose of kitchen left-overs quickly and hygienically. They can be fitted to most sinks and are particularly useful with double-bowl types. Modern sinks with a 90mm outlet will accept the unit directly; older ones with a 38mm outlet need an adaptor.

What to do

First choose your model. Some are batch-feed types while others accept continuous feed, and there are differences in size between individual models so it is wise to check the available clearance under an existing sink.

Follow the manufacturer's instructions carefully to install the unit. The first step is to fit the inlet assembly and plug to the sink outlet, followed by the mounting ring which fits round the sink's waste outlet. Then loosen the mounting bolts on the waste-disposal unit and secure it to the mounting ring. Finally, connect a trap to the unit's outlet and link this to the waste pipe. You may need to make some minor modifications to the exisiting pipe run before you can make the connection.

Provide a power supply for the unit via a switched fused connection unit with a neon on/off indicator, run as a spur from a socket outlet in the kitchen. If the unit has its own on/off switch, site the connection unit in the cupboard under the sink.

What you need:
- waste disposal unit
- trap and waste pipe plus connectors
- plumbing and general-purpose tools
- fused connection unit
- 2.5mm² cable

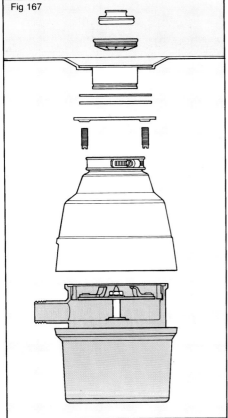

Fig 167

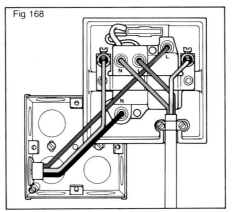

Fig 168

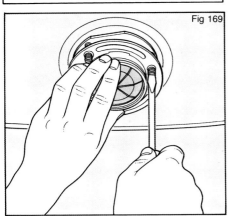

Fig 169

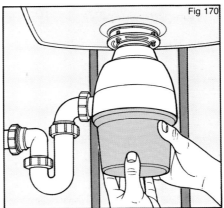

Fig 170

Fig 167 A waste-disposal unit attaches directly to a special mounting ring fitted to the sink outlet. On older sinks, you can either enlarge the hole or fit a special narrow adaptor.

Fig 168 Provide a power supply to the unit via a fused connection unit (FCU) wired as a spur from a nearby socket outlet. If the unit has its own on/off switch, mount it above the work-top and run cable up to it from the FCU, then take it back down to a flex outlet plate next to the unit. Wire the unit's flex into this.

Fig 169 Start by fitting the outlet and mounting ring to the sink waste.

Fig 170 Attach the unit to the mounting ring, fit the trap and connect it up to the waste pipe.

Plumbing in a Washing Machine

You *can* use an automatic washing machine by connecting the inlet hoses to your kitchen taps and hooking the outlet hose over the side of the sink, but it is a great deal more convenient to plumb it in permanently. This involves providing hot and cold supplies (for most machines this is true; however, a few take cold water only) and also some means of getting rid of the waste water. Older machines used to rely on siphonage, which meant the use of an open-ended standpipe and trap, but newer machines now pump out the waste water so you can connect the drain hose to a nearby waste pipe via a special trap, a swept tee or a screw-in connector. Screw-in connectors can also be used to link the supply hoses directly to the kitchen water supplies without the need to turn the water off, and may incorporate a stoptap and a threaded outlet so the machine hoses can be linked directly without the need for any other fittings.

Make sure that any connectors you buy are approved by the Water Research Centre.

What to do

Start by deciding what supplies are needed and how you are going to get rid of the machine's waste water. Use a standpipe and trap if the washing machine manufacturer recommends it, running the waste pipe to a gully or a soil stack. Otherwise either run the waste pipe to a screw-in connector attached to the sink waste pipe, or else change the existing sink trap for one incorporating a special outlet for the machine hose.

You can cut conventional tees into the supply pipes and run each branch pipe to a mini-stoptap, or use a tee with a stoptap as its outlet. However, the quickest method is to use a screw-in connector with a stoptap and threaded outlet. Then all you need do is simply screw the machine hoses to the stoptap outlet. (*see* Fig 172.)

You can plumb in a dishwasher following precisely the same instructions. Note that most dishwashers take only a cold water supply.

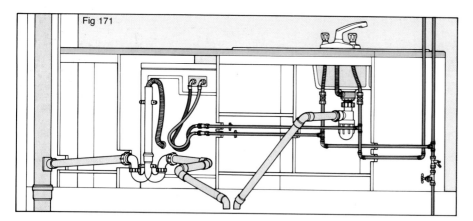

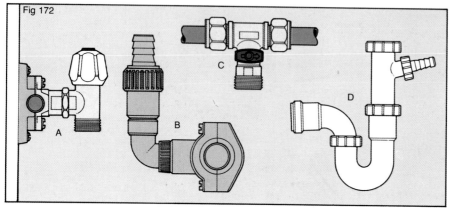

Fig 171 The traditional way of plumbing in a machine, with teed branch pipes providing the water supplies and the waste going into a vertical standpipe.

Fig 172 Self-cutting screw-in connectors can be used for both supply (A) and waste connections (B). As a supply alternative, cut in a tee with integral stoptap (C). Outlet hoses can also be connected to special traps (D).

OUTDOOR JOBS

Plumbing does not have a high glamour rating as far as the great outdoors is concerned. Our use of water in the garden is generally restricted to mundane things like watering the plants and washing the car. Of greater importance are the systems that get rid of the house's waste water and the rain-water that falls on its roof surfaces.

Apart from explaining how to cope with the occasional problems caused by blocked drains (see page 78), this book does not concern itself with what goes on underground; laying new drains is a job best left to the qualified installer. This section deals instead with two popular outdoor jobs: the provision of an outdoor water supply, and the correct planning and installation of gutters and downpipes.

Water out of Doors

An outdoor water supply has to cope with two potential dangers: frost and physical damage. For this reason, supply pipes are buried well underground wherever possible, and need insulating where they emerge above the surface. There is also the problem of corrosion in certain soils, which makes the use of plastic piping a better bet than copper.

The frost problem does not affect rain-water systems, which never run full and spend much of their lives dry. However, these systems do have to be able to withstand continuous exposure to the elements and also the risk of damage from inexpertly wielded ladders. Plastic materials have now largely superseded cast iron, offering major advantages as far as ease of installation and maintenance are concerned, but at the risk of some fragility.

Plastic has also largely taken over for underground drainage work, replacing the brick-built manholes and earthenware pipes of yesteryear. Its main drawback is again its comparative lack of rigidity, which must be compensated for by careful laying techniques – the main reason why drainage work should be left to the professional.

Water for pleasure

Aside from watering the plants, the other use of water in the garden is as a feature – to fill a garden pond, trickle over a man-made waterfall or splash from a fountain. Water gardening is a craft in itself and can add a whole new dimension to your enjoyment of your garden; there is nothing more soothing than the sound of running water, and it also acts as an oasis for wildlife. You may plan a pond as a home for your goldfish, but it will also attract birds, wild animals and insects and will rapidly be colonized by the local frogs, toads and newts.

Creating a pond is child's play; all you need is a hole in the ground and a rigid or flexible liner. Fill it with water, add the essential plants to oxygenate it and you can introduce the fish within a few weeks as soon as the plants have become established. Making the water flow is simple too – whether you want a fountain or a waterfall. You simply install a low-voltage pump run from a transformer to provide the water circulation; you can even add poolside or underwater lights if you wish. A well-established pond will need minimal maintenance; just the occasional top-up of water in hot weather.

Fig 173 (*above*)
Disposing of rain-water and watering the garden are the main features of outdoor plumbing. A rain-water butt copes with both.

Fitting an Outside Tap

One of the most useful plumbing jobs you can do out of doors is to install an outside tap. It can be sited on the house wall, in an outbuilding, or remote from the house – whichever is the most convenient. Once it is installed you will find it much easier to water the garden, wash the car, hose down the patio or tend produce in your greenhouse without the hassle of running hoses through windows to a tap inside the house, or endlessly filling and carrying watering cans.

Whichever site you choose for your tap, the installation is extremely simple, involving teeing a branch supply off the house's main supply pipe at the most convenient point. You can buy all the components you need separately, but if you are planning to position the tap on the back wall of the house and close to the rising main indoors, the quickest way of obtaining everything you need is to buy a complete kit of parts – available from most DIY superstores. All you will need to complete the installation are some basic plumbing and general-purpose tools.

What to do

Start by deciding where you want your new tap to be, and at what point you will make the connection to the main supply pipe since this will dictate the length of the pipe run. Then work out what components you will need to make it up. You can use ordinary copper piping, flexible pipe or plastic pipe; the last is probably the best choice, because of its higher frost resistance.

For a typical installation on the wall of the house, start by fixing the tap to the wall. Then make a hole through the wall for its supply pipe – you need a hole about 22mm (1in) in diameter and you can make it either with a core drill, or by chipping away the masonry with a cold chisel and club hammer. If you are using the latter method and you have cavity walls, drill a pilot hole through both leaves with a long masonry drill first to help ensure that the two holes are in line with each other. When the hole is made, line it with a sleeve of 22mm copper or plastic over-

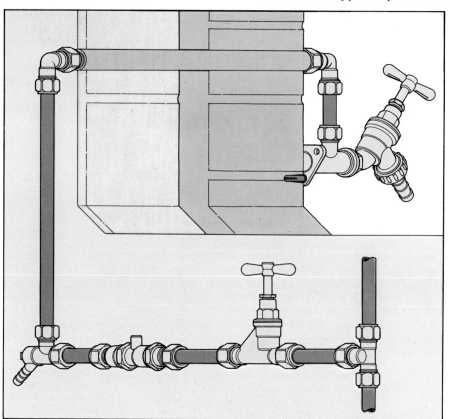

Fig 174   A pipe run to an outside tap should include a stoptap (to allow the run to be isolated), a draincock (so it can be emptied in cold weather) and a double-check valve (to prevent back-siphonage if the mains pressure drops).

flow pipe to protect the new supply pipe.

Fit an elbow to a length of pipe slightly longer than the wall thickness and pass it through the wall so the elbow meets the wall surface. Mark the other end so you can cut it back to leave just enough pipe for the second elbow to be fitted. Then link the outside elbow to the backplate elbow with a short length of pipe, and cover it with pipe insulation and water-proof tape.

On the inside, attach another length of pipe to the inside elbow, long enough to reach down to the level of your connection point to the rising main. Fit a draincock elbow at this point, so you can drain the branch pipe of water in frosty weather – an additional precaution to avoid a burst pipe. From there, make up the rest of the run to the connection point via a double-check valve (to prevent the risk of back-siphonage if there is a drop in mains pressure – a requirement of the new water by-laws) and a stoptap to allow you to isolate the supply. Complete the job either by cutting a tee into the rising main (after turning off the rising main stoptap and draining the pipe) or by using a self-cutting screw-in connector.

Note that in theory you should notify your local water authority of your intention to install an outside tap, although in practice people seldom do. The authority will also want to increase your water rates for the use of a garden hose or a sprinkler.

As an alternative to using ordinary pipe and a pair of elbows to form the run through the house wall, you could use a length of hand-bendable copper pipe for this section instead.

Fig 175 (*above*) An outdoor tap can prove extremely useful.

Remote Taps

If you want to install a tap remote from the house, you will have to run it underground in a trench about 450mm (18in) deep. Use medium-density polyethylene pipe for the whole run from the indoor wall elbow onwards, with purpose-made plastic or brass compression fittings. At the tap position, bring the pipe to the surface within a sleeve of uPVC rain-water pipe, held to the wall or a post with pipe clips, and fill this with foam or granular insulation to protect it from frost. Seal the top with mastic to keep the water out.

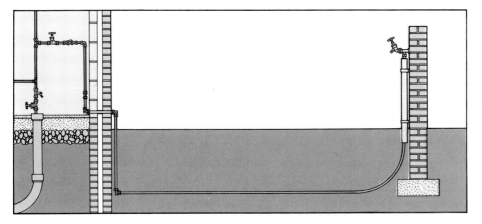

Fig 176 Run plastic pipe underground at a depth of at least 450mm (18in) to serve a remote tap. Protect the above-ground section from frost by running it through a length of downpipe filled with insulation.

Fitting Gutters and Downpipes

The purpose of gutters is to prevent rain-water from cascading off roofs, soaking anyone underneath and causing damp and damage to the house walls. The water they collect is carried to underground drains or soak-aways by vertical down-pipes, which discharge their contents either via an angled spout (called a shoe) over an open gully, or, in most modern homes, directly into a sealed back-inlet gully.

Perimeter gutters and their downpipes were made almost exclusively of cast iron until recent years – plastic has taken its place. Cast-iron rain-water goods are still available, but are now used mainly for authentic restoration work on older homes. The advantages of plastic are that it is light in weight, so is easy to handle and install, needs virtually no maintenance and – as a fringe benefit – is unsafe for burglars to climb. Set against these is its relative weakness – it can easily be damaged by ladders and the like.

What to do

If you are planning to install new gutters and downpipes, it is highly likely that first you will have to remove an existing rain-water system. So start by investigating how it was put together.

Half-round guttering is supported on brackets, which are usually fixed to the fascia at eaves-level, but are sometimes nailed to the sides of the roof rafters. Ogee-shaped guttering has a flat back and is fixed by screws driven through this into the fascia behind. On both types, adjacent lengths are generally bolted together and they may also be bolted to their brackets. Downpipes have interlocking sockets and spigots, generally loose-jointed, and may be secured to the house walls with separate downpipe brackets or by screws driven through integral lugs on the socket ends of each section of pipe.

The problem you have to overcome in dismantling such a system is rust. All the

What you need:
- gutter sections
- gutter unions
- union brackets
- stop-ends
- gutter outlets
- 90° and 135° corner fittings
- gutter brackets
- rain-water hoppers for valley gutters
- downpipe sections
- straight and branch connectors
- offset bends
- downpipe brackets
- shoes or connectors for back inlet gullies
- fixing screws
- wall-plugs
- general-purpose tools

CHECK
- that where a run turns a corner, the correct fall is maintained to the next outlet
- that you leave expansion gaps at gutter joints, as recommended by the manufacturer

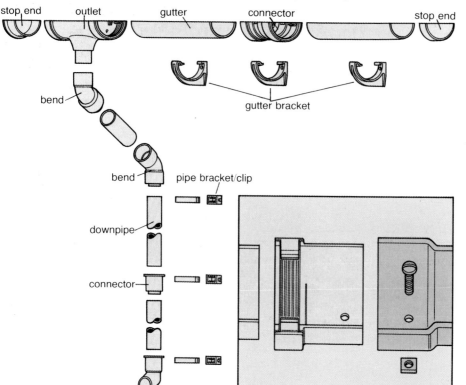

stop end · outlet · gutter · connector · stop end

gutter bracket

bend

bend · pipe bracket/clip

downpipe

connector

Fig 177 The various components of a plastic rain-water system. Special connectors (inset) are available for connecting plastic guttering to cast iron – often necessary on semi-detached properties.

64

Fitting Gutters and Downpipes

fixings are likely to have corroded and there is little point in wasting time trying to undo them. Saw through fixing bolts, and either drill off the heads of rusty screws or chop through them with a sharp cold chisel. Start by disconnecting lengths of gutter; then get a helper to give you a hand in lifting them off their brackets –

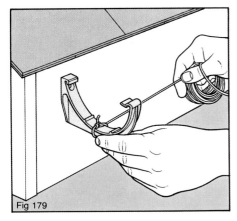

Fig 178 (*above*)   Plastic gutters are easier to handle and install, and need virtually no maintenance.

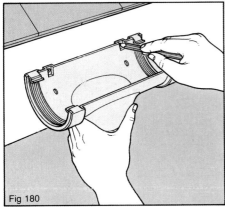

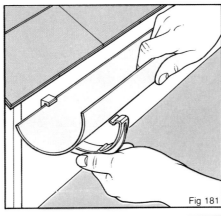

Fig 179   Stretch a string-line from the highest bracket to set the fall on each gutter run.

Fig 180   Drop a plumb-line over gully positions to help position gutter outlets correctly.

Fig 181   Lay each gutter section in the brackets and snap the front of the bracket over the gutter edge.

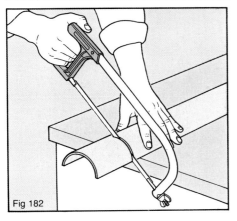

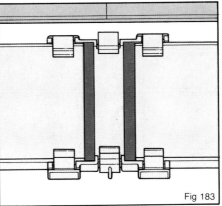

Fig 182   Cut guttering to length where necessary with a hacksaw or other fine-toothed saw.

Fig 183   When clipping lengths together, leave an expansion gap.

65

Fitting Gutters and Downpipes

they are too heavy to manage single-handed. Next, remove all the old gutter brackets and make good the fascia. Lift out offsets linking gutter outlets to their downpipes, then lever out the downpipe fixings so you can lift off each length in turn.

Measure up carefully for your new system – use the old system as a guide to work out what fittings will be needed. For house roofs, you should use 112mm wide guttering with 68mm downpipes; smaller sizes are suitable only for outbuildings or small bay window roofs. Plastic guttering and downpipe comes in 2, 3 and 4m lengths; use the longer lengths wherever possible to avoid joints. Black, grey and white are the most widely available colours, but a few firms also offer dark green or brown as alternatives.

Start the installation by fixing the gutter brackets to the fascia boards. Each run of guttering should have a fall from its closed top end to the point where it discharges into a downpipe; this should be about 5mm ($\frac{1}{5}$in) for every 1m (3ft) of the run. Fix the first bracket at the top end, as high on the fascia as possible, and run a stringline to the other end. Hold it level, then lower it by the required amount and mark the position of the low end bracket. Fix this bracket in place, then add the intermediate brackets at roughly 1m (3ft) intervals.

Fit a stop-end to the first length of gutter and lift it into place. Add other lengths as required, joining them as per the manufacturer's instructions. Use corner fittings as necessary and fit running outlets at downpipe positions. Next, make up the downpipe runs, clipping them to the wall every 2m (6ft) and complete the job by making up the offset section that links the running outlets to the tops of the downpipes. Finally, test the system by pouring buckets of water into it.

TIP 1
Drop a plumb-line from the eaves to a gully at ground level, so you can position the outlet precisely over the gully and ensure that the downpipe will be truly vertical.

TIP 2
To ensure that you cut gutter lengths squarely, use an offcut to mark your cutting line.

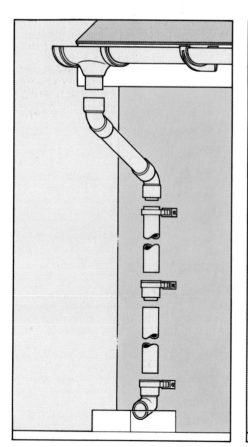
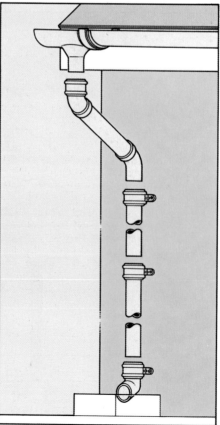

Fig 184 Make up downpipe runs from the bottom upwards, securing each length with a bracket round the pipe connector (far left) or by fixing the socket lugs directly to the wall (left). Complete the job by making up the offset between the top of the pipe and the outlet.

REPAIR JOBS

Our water supply is clean (for the most part), efficient and almost universally available. It does its job so well that we take it completely for granted – until something goes wrong. The trouble may be caused by a fault developing somewhere in your home's plumbing or heating system, the malfunction of an individual appliance, or a blockage. It may be the result of an accident (such as driving a nail through a hidden pipe) or of simple wear and tear. After all, some parts of the system are at work constantly, so it is not unreasonable to expect a breakdown occasionally.

Whatever the problem, it is well worth trying to track the fault down and putting it right yourself. Of course, not everything is within the scope of the do-it-yourself plumber, but many repairs are simple to carry out and doing them yourself will certainly save you money.

What to Tackle

The golden rule when tracing any plumbing fault is to be methodical. If you are, you will usually be able to work out what is causing a particular fault and take some remedial action, even if you are unable to put it right yourself. This will at least put your mind at rest and once you've isolated what is causing the problem you will be able to get everything else working again. If you are able to repair the fault yourself, you will have saved money, time spent in waiting for a repair man to come and also face – it's surprising how many experts report being called out to fix faults that involved nothing more than remaking a leaking connection or replacing a worn tap washer, to the customer's great embarrassment.

However, don't be tempted to try things beyond your capabilities. This point applies particularly to the majority of jobs involving your heating system, and to heaters supplying hot water, which are often quite complicated. The repairs dealt with in this section are of an essentially general nature; leave anything not covered here to an expert. This applies especially to all gas-fired appliances; leave these to the experts.

Emergency Action

The biggest problem with some plumbing faults is the fact that there is water everywhere, ruining decorations, furniture and possessions. The most important thing to do is to stem the tide and minimize the damage caused, and so it is vital that everyone in the family knows where the main stoptap is and how to operate it. The same applies to other controls such as gate valves and mini-service valves on individual water-using appliances. You also need to know how to drain down various parts of your system in an emergency (see pages 32 and 33 for more details). If this involves draining the heating or hot water systems, don't forget to turn off the boiler or water heater first.

It is also a good idea to keep a small kit of tools and materials somewhere handy to cope with emergencies (see page 18).

Who Owns What

The incoming mains supply pipe is your responsibility from the point where it leaves the water authority stoptap. So are drains on your property, although there may be shared responsibility with neighbours for communal drains.

Fig 185 (*above*) Many plumbing faults are relatively straightforward to track down, and are often surprisingly easy to put right yourself.

Curing Dripping Taps

Taps are used several times every day, so problems are bound to develop occasionally. The commonest faults are drips from the tap spout and leaks from the tap spindle. To put things right, you need to know what sort of tap you are dealing with.

Most taps in use in our homes have a spindle mechanism which raises and lowers a small piston called a jumper, on the end of which is a rubber washer. This seals the water inlet inside the tap when it is turned off. Taps may have rising or non-rising spindles – you can tell which by turning the tap on and checking whether the handle rises perceptibly or not.

Some homes still have the unique Supatap with its overhead supply. On these the washer is attached to an anti-splash device inside the rotating tap nozzle.

New installations may have taps with ceramic discs instead of a washer. You can identify such taps by their action; they need only a quarter-turn from off to full on.

What to do

Drips from the spout of a washered tap are usually caused by a worn tap washer, or by a worn inlet (the seat) inside the tap body. To replace a washer, first turn off the water supply to the tap concerned (see page 32 for more details) and open the tap. Remove the tap handle next. On old-style taps with rising spindles, this is usually held on by a small grub screw (see Fig 188). Unscrew the shroud to reveal the tap mechanism. If it is stuck fast, try running some penetrating oil down the spindle, pouring boiling water over it or gripping it with an adjustable wrench after wrapping it in cloth to protect it. On modern taps with non-rising spindles, the handle may simply pull off, or may be held on by a screw underneath the hot or cold indicator disc. Prise this off with a small screwdriver.

Now you can unscrew the tap mechanism – anticlockwise when viewed from above.

MIXER SPOUTS

Kitchen mixer taps may leak round the base of the swivel spout. To cure the fault, remove the spout. Undo a visible grub screw, remove a screw-on shroud and release the circlip under it, or simply pull the spout upwards. Prise off the old O-rings on the spout tail and fit replacements.

SUPATAPS

Supatap sizes have changed, so it you are rewashering an old tap, take the washer with you so you can buy a replacement of the correct size.

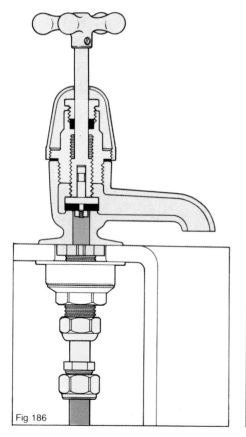

Fig 186

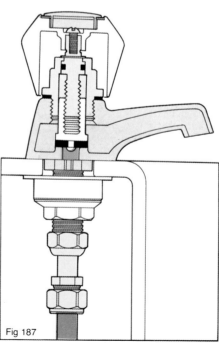

Fig 187

Fig 186 Rising spindle taps have a shroud which conceals the tap mechanism.

Fig 187 Non-rising spindle taps have handles that pull off or are retained by a screw beneath the hot/cold indicator disc.

Curing Dripping Taps

Brace the tap to stop it turning as you do this, and lift the mechanism out to reveal the washer. This may be a push-fit on the jumper, or may be retained by a small screw or nut. Remove it and fit a replacement, then reverse the dismantling procedure to reassemble the tap. Check that the drip has stopped.

With Supataps you do not need to turn the water off. Simply loosen the nut above the nozzle so you can unscrew the nozzle from the tap body. A check valve inside the tap will drop down and stop the water flow. Push the anti-splash device out of the nozzle, prise off the old washer, fit a new one and reassemble the tap.

With ceramic disc taps, you cannot fit new discs. Simply unscrew the tap mechanism as for a spindle tap and fit a complete replacement cartridge, available from the tap manufacturer.

Another cause of dripping taps is a worn seat, and fitting a new washer will not cure the problem (although using a domed Holdtite washer instead of the standard flat one will provide temporary relief). Here the solution is to regrind the seat using a special tap re-seating tool. This has interchangeable cutters for ½in and ¾in taps, and is simply screwed into the tap body once the mechanism has been removed. Rotating the tool's handle grinds away the rough surface of the seat, leaving it shiny and smooth and able to act as a watertight bed for the tap washer.

Leaks from the tap spindle are caused by wear on the spindle seals. Old rising-spindle taps have fibre packing; to stop a leak, undo the gland nut, wind several turns of PTFE tape round the spindle and pack it down with a screwdriver. Replace the gland nut, taking care not to over-tighten it. Non-rising spindle taps have rubber O-ring seals. Prise off the circlip holding the spindle in the tap mechanism. Remove the spindle, prise off the old O-rings and fit replacements of the same size smeared with a little silicone gel to ensure smooth operation. Replace the spindle, circlip and handle.

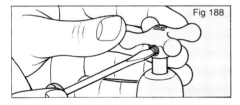

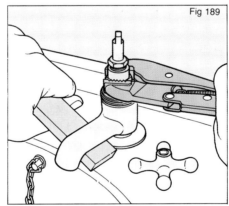

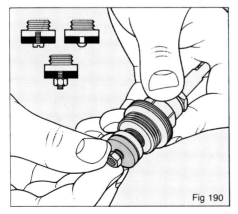

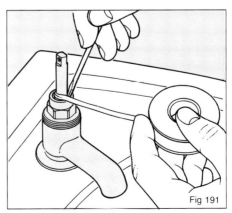

Fig 188   Turn off the water supply and remove the tap handle (after removing the securing screw if necessary).

Fig 189   Brace the tap and unscrew the mechanism from the tap body.

Fig 190   Remove the old washer and fit a replacement.

Fig 191   On rising spindle taps, repack a leaking gland with PTFE tape.

Repairing Leaks and Bursts

The consequences of leaks and burst pipes can be very serious in terms of the damage they can cause to your home. A major burst can literally flood the building, bringing down ceilings and ruining decorations, furnishings and possessions. But the less spectacular problems can be just as bad; the tiny pinhole leak or the regular dripping from a loose fitting can cause rot to develop in structural timbers if it is undetected, and an extensive outbreak can be very expensive to rectify.

Problems with the central heating system are particularly unpleasant because the contents of most systems are not the nice clean water that comes out of the tap, but a disgusting black sludge which will ruin floor coverings and be very difficult to clean up if a leak occurs.

Do your homework now so that you know what to do if you ever suffer a serious leak or burst. Make sure you know how your system works, and be prepared for action when something goes wrong. A prompt response can save a lot of expensive repair work later.

What to do

Before disaster strikes, make sure you know where all the stoptaps, gate valves and draincocks are on your system. Check taps and valves every six months or so, to make sure they haven't jammed open (see page 74 for how to free them). Lastly, keep a small emergency tool-kit (see page 18) so you can at least make some temporary repairs.

If you discover a burst pipe and are in the house on your own, don't panic. Turn off the water supply to the affected pipe, and drain it down as quickly as possible. This means turning off the rising main stoptap for bursts on any mains-pressure pipework – to tanks in the loft, the kitchen cold tap, your washing machine or an outside tap, for example. For bursts on other supply pipes, open all the appropriate hot or cold taps to empty the cold water storage tank, and either turn off the rising main stoptap to prevent it from refilling, or tie up the ball valve to a piece of wood spanning the tank. Then switch

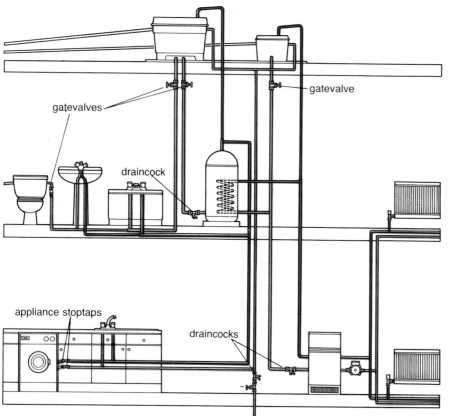

gatevalve

gatevalves

draincock

appliance stoptaps

draincocks

Fig 192 Make sure you know where all your stoptaps, gate valves and draincocks are before disaster strikes. It pays to label them to indicate what they control.

Repairing Leaks and Bursts

off the boiler, immersion heater or other water heater. If the leak is near any electric wiring, turn off the power at the house's main isolating switch. Try to contain the leak with bowls or absorbent cloths – or at a pinch by the time-honoured method of putting your finger over the hole – until the flow stops.

For leaks on the heating system, you should also switch off the boiler. Then turn off the stoptap on the supply to the feed-and-expansion tank, attach a hose to the lowest draincock on the system and lead it to a gully or drain outside. Open the valve to empty the system. Also see Tip 2.

If you caused the burst by driving a nail or screw into a pipe, leave it where it is while you turn off the water. If you drill into a pipe, drive a screw into the hole to staunch the flow.

Once you have things under control, you can attempt a repair. If the burst is small and in the middle of a pipe run, cut through it at the burst site and remove about 20mm of pipe so you can reconnect the cut ends with a compression fitting (see Fig 193). If the split is more extensive, cut out the damaged section and fit a length of new pipe (see Fig 194); plastic pipe and fittings are ideal for this (but remember you must maintain earth continuity). If you don't have any appropriate fittings available, use PVC insulating tape and a section of garden hose and some strong wire to make a temporary repair (see Fig 195) – at least then you can restore the water supply.

You may be able to cure leaky compression fittings either by tightening up the cap nut, or by opening the fitting and winding PTFE tape over the olive before reassembling it. With capillary fittings it is usually impossible to re-solder a leaking joint because the pipe will contain some water. Instead dismantle it and fit a compression joint.

TIP 2
If a radiator starts to leak (usually as the result of internal corrosion), turn off its isolating valves, open the air vent and either place a container under the leak or undo the valve couplings so you can drain it completely, ready for replacement.

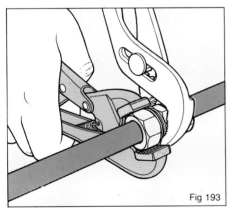

Fig 193

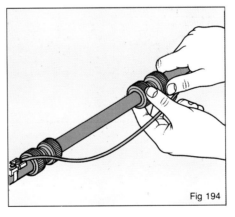

Fig 194

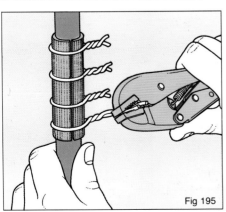

Fig 195

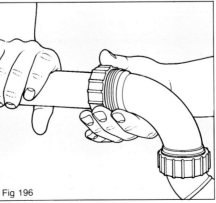

Fig 196

Fig 193   Cut in a compression fitting to repair a small burst.

Fig 194   To repair larger-scale damage, cut out the affected section of pipe and insert a new length. If using plastic pipe, remember to maintain earth continuity with a strapping wire and pipe clamps.

Fig 195   Make a temporary repair with PVC tape, a section of garden hose and strong wire.

Fig 196   If a waste pipe or trap develops a leak, try tightening the connections. If this fails open the fitting, inspect the sealing washer and replace if necessary.

Curing Overflows

As with taps, the ball valves controlling waterflow into storage cisterns, feed-and-expansion tanks and WC cisterns get a lot of wear and tear, so it is not surprising that they malfunction from time to time – few householders have escaped discovering water dripping relentlessly from an overflow pipe, usually in the dead of night! Such problems should not be ignored since the drip may turn into a flood if the valve fails completely, and the overflow might not be able to cope with the volume; it is intended only as a *warning* pipe. Overflows can also cause problems with penetrating damp.

What to do

Before blaming the valve, check the float. If it is partly submerged it has probably sprung a leak and filled with water. Tie up the valve arm to a batten laid across the cistern to stop it filling with water, unscrew the float and buy a matching replacement.

If this is not the cause of the problem, the fault probably lies within the valve. Old brass types such as the Portsmouth valve (*see* Fig 197) contain a piston with a washer on the end, and this wears out. Turn off the water supply and dismantle the valve using pliers to remove the split pin holding the float arm, then unscrew the end cap and prise out the piston. Hold the piston with a wrench and use another to unscrew the washer cover, then prise out the old washer and fit a replacement. Clean the valve out thoroughly before reassembly.

More modern valves contain a diaphragm instead of a washer (*see* Fig 198) and are generally far more reliable. If they give trouble, the diaphragm usually needs replacing. This involves unscrewing the retaining nut, removing the end cap and plunger and prising out the old diaphragm with a small screwdriver. Note which way it was fitted and insert the new one the same way round, then reassemble the valve. The sequence of operations is similar for all diaphragm-type valves.

What you need:
- replacement float
- pliers
- adjustable wrenches
- screwdriver
- replacement washer
- replacement diaphragm

VALVE TYPES
See Fig 199, below.

A: Portsmouth valve.
B: Torbeck diaphragm valve.
C: New brass diaphragm valve to BS1212.
D: Garston diaphragm valve.

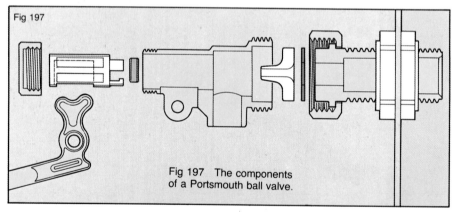

Fig 197 The components of a Portsmouth ball valve.

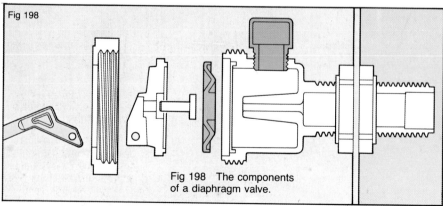

Fig 198 The components of a diaphragm valve.

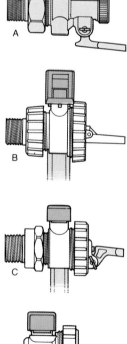

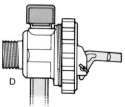

Fig 199

Curing Airlocks and Noisy Plumbing

An airlock is caused by air getting into the plumbing (or occasionally the heating) pipework, and preventing the water from flowing as it should. It most commonly affects hot taps; the symptoms are poor water flow accompanied by hisses and bubbles, then no flow at all.

The most likely cause of an airlock is when the system has been partially drained for repairs or alterations, but it can also be caused by drawing off a lot of water at once from various parts of the system, resulting in the cold water storage tank being emptied and air entering the system via the feed pipes running from the tank.

What to do

Professional plumbers with strong lungs regularly blow airlocks out, but it is easier to use mains-pressure water instead. To cure an airlocked tap, connect one end of a garden hose to it with a hose connector and take the other end to the kitchen cold tap. Open the affected tap and then the mains-pressure one in order to blow the airlock out. *See also* Tip 1.

If the airlock is affecting a Supatap, unscrew and remove the nozzle and connect the hose direct to the tap outlet. If it affects a kitchen mixer tap, remove the swivel spout and hold a cloth firmly over the spout opening. Turn on the hot tap first, then the cold one. Turn off the taps when the airlock is cleared and *before* you remove the cloth.

If you get airlocks regularly in your water supply pipework, there are several points worth checking. You may need a bigger cold water storage tank, especially if you have added a lot of new water-using appliances in an old house. You may need to clean or replace the ball valve on the storage tank if it is sluggish and cannot fill the tank as fast as water is drawn from it. Lastly, the pipe from the storage tank to the hot cylinder may be too small (it should be 22mm in diameter) or its gate valve may be partially closed; this will cause the water level in the vent pipe to fall below that in the hot cylinder as hot water is drawn off, resulting in water being drawn into the pipes.

See page 81 for information on curing radiator airlocks.

Curing Noisy Plumbing

Pipes carrying hot water expand as they heat up and contract as they cool. This can cause ticking or creaking noises which are very irritating. Check points where pipes pass through holes in floors, joists and walls, and pack pipe insulation round them. Do this too where pipe runs are held rigidly by pipe clips, and remove any clips close to right-angled bends to allow the pipe to expand freely on either side of the bend.

Humming in pipes may be due to undersized pipe being used, or because the pump speed is set too high on heating systems. If you have this problem, try reducing the pump speed slightly (see page 83); call an expert in if the noise persists.

Cure noise from filling cisterns by replacing the ball valve with a new, quieter type (see page 72).

Banging sounds emanating from the boiler may be due to a build-up of scale, or to air being drawn into the system. (See page 82 for more details).

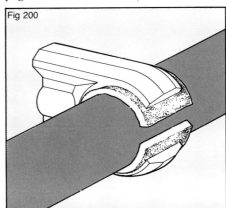
Fig 200

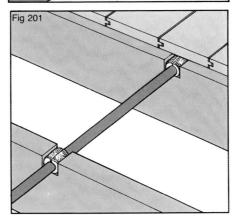
Fig 201

Fig 200   Make sure pipe runs are supported by clips, with padding added to reduce expansion noise.

Fig 201   Pack round points where pipes pass through joists and walls, using pipe insulation or glass-fibre insulation pads.

What you need:
For clearing airlocks
- garden hose
- hose connectors
- back-flow check-valve

For noisy pipes
- pipe clips
- pipe insulation

TIP 1
The new water by-laws require any connection – even a temporary one such as the hose described here – between the mains and stored water to be protected against back-siphonage by the inclusion of a double-check valve. You can buy special hosepipe connectors which incorporate back-flow protection devices, and it's wise to keep one in your tool-kit.

TIP 2
Avoid airlocks on new pipe runs by ensuring there is always a slight fall away from the main vent pipe. If 'high spots' are unavoidable, fit an automatic air-release valve at the highest point on the pipe run.

Freeing Stoptaps and Ball Valves

One of the most serious plumbing problems is to find that you cannot shut off a stoptap or gate valve in an emergency. If this happens to you, the first thing to do is to try to contain the water flow (*see* pages 70 and 71 for more details) to minimize the damage. Turn off the outside stoptap too if you know where it is and have a key to turn it. Then you can turn your attention to freeing the jammed tap or valve in the house.

A ball valve that is jammed in the open or closed position can also be a serious problem. Jammed open, it could cause the tank or cistern concerned to overflow (remember that the overflow pipe is really just a warning pipe designed to warn of a leaking valve, and it may not be able to cope with the flow if the valve is fully open). Jammed closed, the tank or cistern will run dry and this will cause airlocks in the plumbing pipework, and possible damage or overheating of the central heating system.

What to do

With a jammed stoptap, the first method to try is trickling some penetrating oil down the tap spindle; leave it for a few minutes, then try to operate the tap. If this fails, try improving your leverage by using a wrench on the tap handle. However, this can be dangerous; do not apply too much force, or you might shear the handle off completely. Method three is to apply heat to the tap or valve body using a blowlamp or hot air gun; the resulting expansion of the metal parts may free them. Remember to let the tap body cool down before trying to operate the handle. If all else fails to shift a jammed stoptap, turn off the outside stoptap (*see* page 33) and then replace the jammed tap. On low-pressure pipework, drain the pipe run concerned and fit a new gate valve.

With a jammed ball valve, turn off the water supply to it and remove it so you can clean or replace it (*see* page 72).

What you need:
- penetrating oil
- blow lamp or hot air gun
- new stoptap or gate valve
- new ball valve
- plumbing tools

CHECK
- that gate valves are left fully open
- that you fit a replacement stoptap the right way round, with the arrow on its body pointing in the direction of the water flow. Gate valves work either way round
- that a replacement ball valve matches the supply pressure. You may need to fit or remove a special nozzle depending on whether the supply is at high or low pressure

TIP
Make sure that your stoptaps and gate valves stay 'free' by operating them regularly.

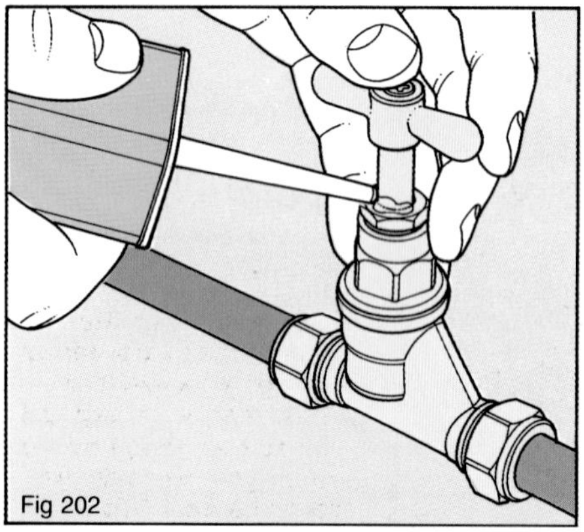

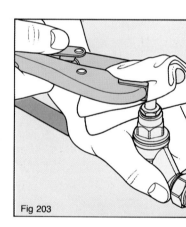

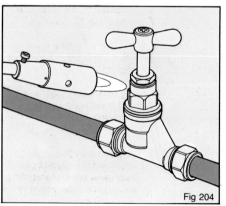

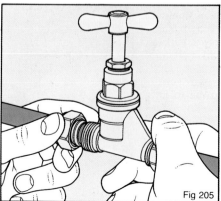

Fig 202   First, try to free a jammed stoptap by trickling penetrating oil down the spindle.

Fig 203   Next, use a wrench to exert extra leverage on the tap handle. Take care not to break the handle or shank.

Fig 204   If that fails, try applying heat to the tap to free the mechanism.

Fig 205   Lastly, turn off the outside stoptap and replace the jammed tap.

Curing WC Flushing Problems

If your WC cistern will not flush properly, the diaphragm inside the siphon mechanism has probably torn or perished as a result of years of everyday operation, and will need replacing. This is a relatively simple job and the new diaphragm will restore the flushing action perfectly.

There are two other faults which may also affect the WC cistern. The first is a broken or uncoupled lever mechanism, allowing the handle to move freely without operating the siphon. The second is perpetual flushing, where water continues to run into the pan long after the cistern has been flushed.

What to do

To replace a diaphragm, the first step is to turn off the water supply to the cistern and to flush it. Mop out any remaining water with a sponge. If you have a separate cistern, undo the nut connecting the flush pipe to the cistern outlet and then use an adjustable wrench to undo the backnut securing the siphon outlet underneath the cistern. With a close-coupled cistern, disconnect the water supply and overflow pipes, remove any wall-fixing screws, undo the nuts securing the cistern to the back of the pan and lift it off. Remove the siphon nut as before.

Uncouple the flush lever mechanism and lift the siphon unit out (after removing the ball valve arm if necessary). Pull out the lift rod and remove the washers on it, noting their order. Remove the worn diaphragm and fit the replacement. If you cannot obtain an exact match, buy the next largest size and cut it down with scissors. Reassemble the flushing mechanism, replace the siphon and reconnect everything.

To fix a broken or uncoupled lever mechanism simply reconnect it or replace the missing C-ring. To cure perpetual flushing, slow down the flow rate into the cistern by fitting a high-pressure seat inside the ball valve. This will stop the siphon from operating continuously.

What you need:
- replacement diaphragm (also known as a siphon washer)
- replacement C-ring
- high-pressure ball valve seat
- plumbing tools

CHECK
- that your cistern is filling to the correct level, and adjust the ball valve arm if necessary. A partly-filled cistern will not flush properly.

TIP
You can improvise a temporary diaphragm repair by cutting a replacement from heavy-duty polythene sheet.

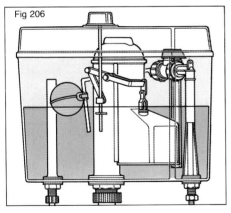

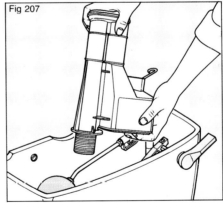

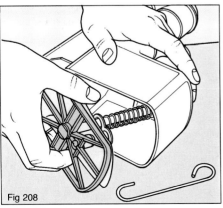

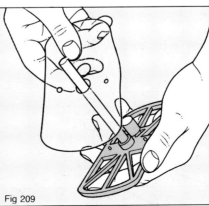

Fig 206   With a close-coupled cistern, disconnect the supply and overflow pipes, undo any fixings and lift the cistern away from the pan so you can undo the siphon backnut. With a separate cistern, simply disconnect the flush pipe and undo the backnut.

Fig 207   Uncouple the lever mechanism and lift out the siphon unit.

Fig 208   Pull out the lift rod, remove the washers and slide off the old diaphragm.

Fig 209   Fit the replacement diaphragm and reverse the dismantling sequence to reassemble everything.

Replacing an Immersion Heater

Most homes have an immersion heater fitted to the hot water cylinder, as back-up for a boiler-heated supply (or instead of it in summer months when central heating is not required). The heater element will eventually fail – quite quickly in hard water areas, where scale forms on the element and makes it work harder to heat the water, or where the water is aggressive (see Tip) – but fitting a replacement is quite straightforward.

Start by checking what type of heater is already fitted. The most likely type is a long single element, but you may have a twin-element heater (wired to a switch with two positions, often marked 'Sink' and 'Bath'), or two separate elements mounted in the side of the cylinder rather than at the top. Buy a suitable replacement (all are designed to fit a standard 2¼in diameter BSP-threaded boss) and also buy or hire a special immersion heater spanner to undo the old heater from the cylinder boss so you can remove it and fit the new one of the appropriate type and rating.

What to do

Start by turning off the water supply to the hot cylinder (or tie up the ball valve and drain the cold tank if there is no gate valve fitted) and run the hot taps until the water stops. Then attach a length of garden hose to the draincock at the bottom of the cylinder, and draw off enough water to lower the water level to below that of the boss.

Next, isolate the power supply to the heater switch and disconnect the flex from the heater. Use your immersion heater spanner to unscrew the old heater; if it won't budge, use a hot air gun to soften the sealant round it. Lift it out and clean off the remains of the old sealant from around the boss.

Slide the new heater in after fitting the fibre washer round it, screw it up hand-tight and give it half a turn with the spanner. Insert the thermostat if it is separate, reconnect the flex and fit the cover. Finally, restore the water supply.

What you need:

- new immersion heater
- immersion heater spanner
- hot air gun
- electrical screwdriver
- garden hose
- draincock key

CHECK

- that the wiring is intact before replacing a non-working heater
- that the thermostat is working; fit a new one if it isn't
- whether the thermal cut-out fitted to some heaters has tripped off, and reset it if it has

TIP

If you live in an area where the water is 'aggressive' – your water company will tell you – buy a corrosion-resistant heater which will last longer than an ordinary one.

Fig 210

Fig 211

Fig 212

Fig 213

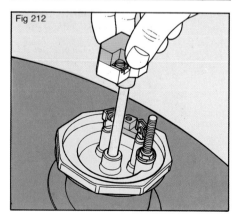

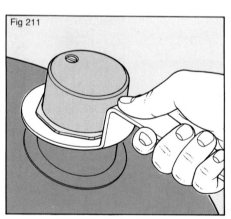

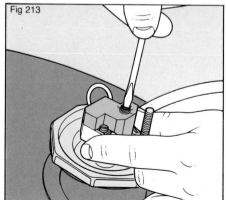

Fig 210 Turn off the water and electricity supply. Disconnect the heater flex and drain the cylinder to below boss level via its draincock.

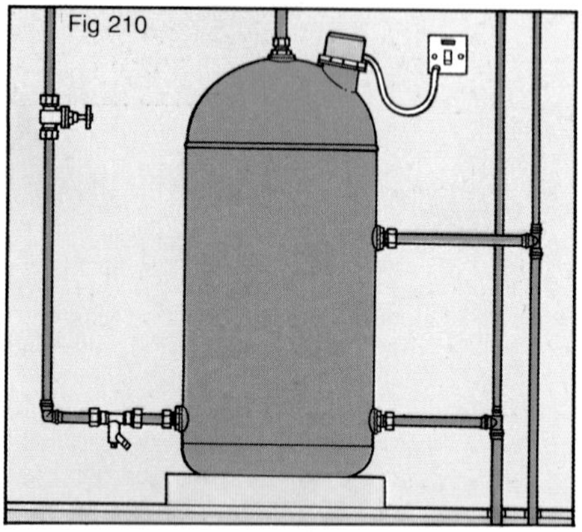

Fig 211 Unscrew the old heater with the special spanner. Don't use too much force or you may buckle the cylinder.

Fig 212 Screw in the new heater and fit the thermostat.

Fig 213 Reconnect the flex and set the regulator to 60–65°C (140 –149°F).

Clearing Blocked Waste Pipes

Although modern plastic waste pipes and traps have smoother inside surfaces than old-style metal ones, they can still become blocked as time goes by. The problem is caused by the things we try to wash down the plug-hole. Tea leaves, hot fat, food particles and household repair fillers are the prime causes of kitchen sink blockages, while hair has blocked many a basin, bath or shower tray trap. The blockage may build up slowly, with water taking longer and longer to flow away as the trap or pipe bore is gradually reduced, and this is a sure sign that trouble is brewing. It's wise to attend to it before the blockage becomes total and you are left with an appliance full of water to contend with as well.

The other problem area is the soil pipe taking waste from a WC. Here blockages may be caused by attempts to flush things like disposable nappies down the pan, or even by such unlikely things as teddy bears going on pot-holing expeditions!

What to do

To clear a blocked waste outlet or WC, start by trying a plunger or force cup. Block sink or basin overflows with a wet cloth first of all, then place the plunger over the waste outlet and plunge it up and down a few times. With WCs you need a larger plunger; place it in the pan and plunge firmly downwards.

If this fails, you must either clear the blockage chemically or physically. Drain cleaners and caustic soda may take a long time to work and must be used with care. Mechanical devices are quicker, but you will probably have to undo some connections to get at the site of the blockage.

Start by replacing the appliance plug. Put a bucket under the trap and unscrew it so you can remove it and clean it out. If the blockage is further down the waste pipe, undo connections so you can introduce a plumber's snake or similar flexible probe to clear the blockage.

Fig 214

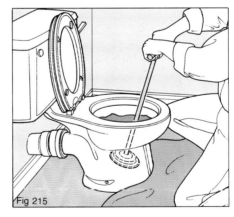

Fig 215

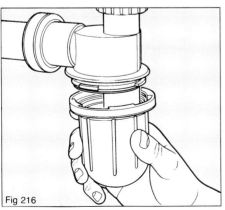

Fig 216

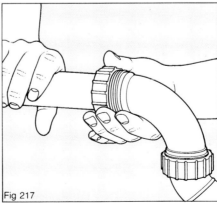

Fig 217

Fig 214 Cover the overflow and use a plunger in a blocked sink or basin.

Fig 215 Use a larger WC plunger to clear blockages between pan and soil pipe.

Fig 216 Unscrew traps to clear blockages below the waste outlet.

Fig 217 Disconnect pipe fittings to gain access to blocked waste pipe runs, so you can insert a plumber's snake or similar flexible probe.

Clearing Blocked Drains

Most people panic at the thought of blocked drains, yet clearing a blockage is generally not very difficult, nor need it be a particularly unpleasant job unless the drain run is totally blocked and is actually overflowing. If that occurs, prompt action is required to minimize the mess that such an overflow can cause.

Although drains run underground, you do not need to know where they go so long as you have access to the various manholes or inspection chambers which should exist at every point where a drain run changes direction, or where a branch drain is connected to the main run. Older homes will have brick-built rectangular inspection chambers with cast-iron or galvanized steel covers, while newer homes will usually have round plastic chambers, again with iron or steel covers. If you cannot locate any of your manholes, suspect that paving has been laid over them at some time in the past, and lift it along the likely line of the run.

What to do

Start by lifting the cover on the manhole nearest the house. If it has rusted in or its handles are broken, tap round the edge with a hammer or apply heat to it with a blowlamp to loosen the seal. Then insert a garden spade under one edge to raise it, insert scrap wood under the edge to support it and lift it aside. Get help with large, heavy covers.

If the manhole is clear, the blockage is between it and the house. Rod the drain in that direction (see Tip). If it is full, the blockage is further down the run and you should lift the next manhole cover to see whether the blockage is between these two, or is further on. If the lowest manhole is full, the blockage is between it and the sewer, and if this is an interceptor trap the cap of the rodding eye may have come off and fallen into the trap. If you suspect this, call in a professional drain-clearing firm to retrieve it.

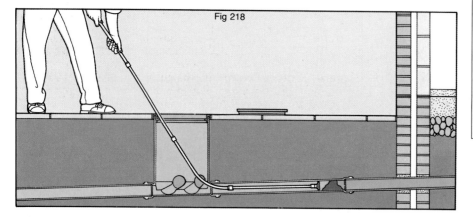

Shared Drains

Most homes have a private drain run to the sewer, and this whole run is the responsibility of the householder including any section outside the property. Drains shared by two or more properties and built before 1937 are generally the responsibility of the local council, although they may pass on the costs of maintenance or repairs to the householders sharing the drains. Shared drains built after 1937 are the joint responsibility of the various households connected to them, so any repair bills should also be shared.

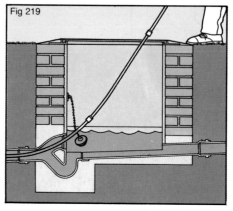

Fig 218 Rod towards the house if the first manhole is clear. For blockages elsewhere you can rod either upstream or downstream.

Fig 219 Rod towards the sewer if all the manholes are full. If you cannot clear the blockage, call in a professional drain-clearing firm.

Clearing Blocked Gutters, Downpipes and Gullies

Blockages may occur in your house's rain-water system as a result of wind-blown and other debris collecting in gutters and being washed into the downpipes and ground-level gullies. The blockage may show up as an overflow at eaves level, as water appearing at downpipe sockets, or as an overflowing gully, and should be cleared at the earliest possible opportunity to avoid the risk of the overflow causing damp penetration through the house walls or other damage to the structure.

Overflows may also occur as a result of defects in the rain-water system – sagging gutter sections caused by failed supporting brackets, non-waterproof gutter joints, or physical damage to gutters and downpipes (commonly caused by carelessly-wielded ladders). By carrying out regular inspection and maintenance of the system, you can spot and put right defects of this sort and also ensure that blockages do not begin to build up unseen in gutters, hoppers and gullies.

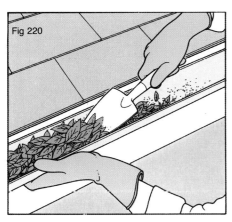

Fig 220

What to do

Gutters   These usually become blocked at bends, at outlets and at low points. To clear the blockage, set up your ladder with a stand-off to hold the top clear of the eaves, and rope it to the building so it is safe and secure to climb. Carry or haul up a bucket and a garden trowel or similar tool, and scoop out as much of the debris as possible, depositing it in your bucket. Then take up the end of your garden hose, and rinse the entire gutter run through with clean water from its highest end to the outlet at the downpipe.

Downpipes   These may become blocked at any point by debris washed in from the gutters; birds' nests are particularly notorious. Start by disconnecting the pipe from the gutter outlet, and use drain rods or a similar long flexible probe to push the blockage downwards. You may have to take down sections of the pipe, working from the top, if you cannot reach the blockage in any other way. Again, rinse the pipe through with clean water once you have cleared the blockage and reassembled the pipe.

Rain-water gullies   These contain a trap underground, and can be blocked both by material washed down from the roof and by debris blown into the gully. Scoop the debris out with a gloved hand, scrub the gully sides and rinse through with water. Fit a gully grid if there isn't one, and if the downpipe discharges below grid level add a gully cover to stop debris being blown in. If the downpipe runs directly into a back-inlet gully, you will have to undo the collar connecting the pipe to the gully to enable you to gain access to the trap.

What you need:

- bucket
- trowel or other scoop
- garden hose
- drain rods
- waterproof gloves

CHECK

- that gutters have a steady fall towards their outlets
- that gutter joints are intact; seal leaky joints on old cast-iron or asbestos gutters with self-adhesive flashing tape, and fit new rubber seals to plastic gutters
- that you do not lean ladders against plastic gutters; your weight may crack them

TIP

Fit a wire cage at each gutter outlet to prevent debris from being washed or blown into the downpipe, and add grids to any unprotected gullies.

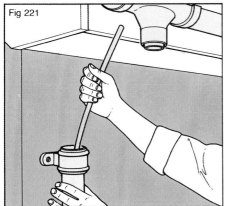

Fig 221

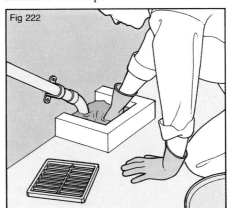

Fig 222

Fig 220   Use a garden trowel or similar implement to scoop debris out of gutters, then wash down the run with water.

Fig 221   Disconnect a blocked downpipe from its gutter outlet and push out the blockage with drain rods.

Fig 222   Clear blocked gully traps by scooping out the debris with a gloved hand. Scrub the sides and wash through with clean water.

Curing Cold Radiators

On a well-designed and properly-balanced wet central heating system, all the radiators should be at approximately the same temperature. However, a variety of problems can occur which will result in one or more radiators on the system being lukewarm or even stone cold.

What to do

When you notice that a radiator is cold, or is not as hot as it should be, start by checking whether this is the only one affected, or whether others are cold too.

One radiator cold Start by checking that the radiator valves are turned on and are working properly. Open the hand-wheel valve at one end of the radiator, and remove the cap from the lock-shield valve at the other end. Use pliers or a small spanner to turn the spindle to full on, noting how many turns this took so you can then restore it to its original position later.

Next, suspect a build-up within the radiator of air or the gaseous by-products of corrosion in the system – particularly likely with radiators at the top of the system. A partially filled radiator suffering from this problem will feel hot at the bottom and cool at the top. Open the air vent with a radiator key to see if air is forced out; if this is the cause of the problem, hot water will run into the radiator and displace the trapped air. Close the vent as soon as water starts to emerge, and check that the feed-and-expansion tank still has water in it.

The third possible cause of a cold radiator is a build-up of sludge – another corrosion by-product – which can block the inlet or outlet to the radiator. Sludge can also cause cold spots along the bottom of the radiator without completely blocking the flow. To get rid of it you usually need to remove the radiator (see page 81) so that the sludge can be completely removed.

Several radiators cold This is most likely to affect upstairs radiators if the heating system is starved of water, due to losses not being replaced via the feed-and-expansion tank. If this tank contains no water (it should be about one-third full when the system is on), check the operation of its ball valve and repair it if it is jammed closed (see page 74). The problem could also be caused by a blockage in the circulating pipes, caused again by sludge. Try increasing the pump speed for a short while to shift the blockage, or drain down the system so it can be flushed through and refilled (see page 83). Lastly, on systems with zone control, the problem may be caused by a faulty motorized valve which is isolating part of the system. It is generally best to leave the servicing and replacement of these components to a professional plumber.

All radiators cold The most likely cause here is a jammed or failed pump (see page 82), or a faulty motorized valve which is diverting water from the heating system. It's worth trying to free a jammed pump (or to replace a failed one) yourself. If the problem is a faulty valve, again call in expert help.

All radiators and water cold Here the fault probably lies with the boiler or programmer. Check that the boiler is alight and the programmer has a power supply. If there appears to be no fault with either of these two components, and the problem persists, call in an expert to find and rectify the fault.

What you need:
- radiator key
- plumbing tools

CHECK
- that radiator valves are open
- that thermostatic radiator valves are operating correctly
- that air is not collecting inside radiators
- that the pump is running properly
- that the system controls are operating
- that the boiler is alight

TIP
Some types of motorized valve cannot be serviced and have to be replaced (a job that will entail draining down the entire system), but on many newer models it is possible to replace the valve motor and free the valve spindle if it is jammed, without having to interfere with the plumbing.

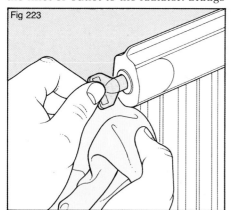

Fig 223 Bleed trapped air from a cold radiator by opening the air vent at the top corner. Close it again as water emerges.

Fig 224 You may be able to drain sludge from a blocked radiator at ground level via a nearby draincock. Fit a hose to it, open the draincock with a spanner and run off some water into a bucket.

Replacing a Radiator

Most radiators used in domestic heating systems are made from pressed steel, and electrolytic action due to the presence of other metals causes corrosion of the system from within. This affects radiators in particular, leading to pinholes and leaks along welded seams, and the resulting escape of water and brown or black sludge – one of the by-products of the corrosion process – can cause a fearful mess. It may be possible to effect a temporary repair by using a two-part epoxy putty, but in the long run the only safe cure is to replace the radiator.

Choose a new radiator that is the closest possible match to the existing one in terms of height and length; new radiators come in metric sizes, and if the one you are replacing is an old imperial size, go for one fractionally smaller than the original if possible. It is easier to fit an extension to the inlet of a narrower radiator in order to reconnect the pipes than to alter the pipework to cope with a wider one.

What to do

Once you have all the necessary materials and tools to hand, the first step is to isolate and disconnect the radiator. Close the handwheel valve at one end, and remove the cover of the lock-shield valve at the other end so you can close this too by turning its spindle with pliers or a small spanner. Note the number of turns this takes so you can repeat the setting once the new radiator is connected and can maintain the balance of flow through it.

Next, position a shallow container under each valve connection, and loosen one of the couplings with a spanner. Brace the valve body against the wall as you do this to prevent it turning and fracturing the supply pipe below. Undo the coupling slightly so water can trickle into the container, and open the air vent to increase the flow rate. Do up the coupling again whenever you need to empty the container. When the radiator is empty, undo the other coupling and stuff tissue paper or rag into the radiator inlets to stop sludge dripping out as you lift the radiator off its brackets and carry it out. Remove the old radiator brackets (the new one is highly unlikely to fit) and measure the back of the new radiator so you can fit the new brackets in the correct position on the wall. Fit the air vent and blanking plug to the top inlet tappings of the new radiator, and screw a valve tail into each of the bottom tappings; you may need a special large hexagonal key called a radiator spanner for this. Wrap PTFE tape round each fitting before screwing it home.

Next, hang the radiator on its brackets and check that the old valves can be connected to the valve tails. Add extension pieces if necessary to bridge any gap. Then slide the coupling nut along each valve tail and connect up to the valves. Open the valves to allow water to enter the radiator, with the air vent open to expel the air. When the radiator is full, reset the lock-shield valve to its original setting and replace its cover.

If you have the misfortune to fracture either of the pipe tails as you try to undo the radiator couplings, you will have to drain down the system to below the level of the radiator you are working on (*see* page 83) so you can repair the damage. Cut out the damaged section of pipe at a point beneath the floorboards and join on a new length of pipe using a compression fitting.

What you need:
- new radiator
- radiator brackets
- screws and wall-plugs
- valve tails
- air vent
- blanking-off plug
- extension connectors
- PTFE tape
- radiator spanner
- radiator key
- shallow containers
- plumbing tools
- general-purpose tools

CHECK
- that you hang the radiator with a slight rise towards the end at which the air vent is situated

TIP
You can use a special Remrad valve to avoid the messy business of draining down a full radiator. This allows you to connect a bicycle pump to the air vent and force the water out of the radiator and into the system pipework before closing the inlet valves and removing the now empty radiator.

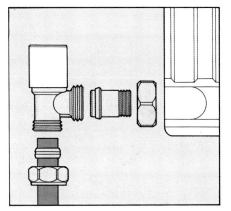
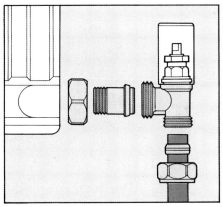

Fig 225 You need to fit a blanking-off plug, an air vent and two valve tails to the new radiator before hanging it on its wall brackets and reconnecting the old valves to it.

Curing Boiler and Pump Problems

Generally speaking, it is wise to leave maintenance and repairs to central heating boilers to an expert service engineer, and sensible householders have their system serviced at least annually to keep it in good condition. However, there are some simple faults with gas boilers (by far the most common type installed) which you may be able to diagnose yourself and so assist the service engineer in selecting the right parts to bring along. You may be able to fix simple pump faults yourself without having to call an engineer in.

What to do

One of the most common boiler faults concerns the operation of the pilot light. This burns continuously on most boilers and lights the main burners when the programmer switches on the main gas supply. If the pilot light goes out for any reason, a heat-sensing thermocouple detects the fault and prevents the main gas supply from operating.

You will usually find the pilot light at the front of the boiler, next to the burners, and it should be visible through an inspection hole. If it is out, try relighting it with a match. No flame suggests a blocked jet, which needs cleaning and adjusting, or a failed thermocouple which needs replacing. On boilers with piezo-electric ignition, simply push the ignition button; if it won't work but you can light the flame by hand, the ignition device is faulty and again needs replacing.

If the system temperature seems erratic, the fault may lie with the boiler thermostat. Try altering the setting on the thermostat dial; if nothing changes the thermostat is probably faulty.

All these parts are simple for an engineer to replace; when you call the engineer out, describe what you think the fault is, and give the make and model number of your boiler if possible.

As far as pumps are concerned, the most common problem is jamming, either after a period when the pump has not been used, or due to sludge in the system jamming the impeller. Fortunately, many types of pump can be restarted without having to be disconnected and dismantled; they have a slot on the end of the shaft, and by turning this with a screwdriver you can generally free the impeller inside. Turn off the pump's power supply before you do this. A little water may seep out from the slot as you free the impeller, but this is no cause for concern.

If you cannot free the impeller by this method, you will have to remove the pump for cleaning. It should have isolating valves on the pipework at either side; Turn these off, disconnect its power supply (with the circuit isolated from the mains) and undo the couplings so you can lift the pump out. Take it to the sink and run water through it while trying to rotate the impeller. When it is free, replace the pump and bleed air from the pump chamber via the bleed screw before restoring the power supply. If the pump still will not run, or it is noisy – a sure sign of worn bearings – it is probably in need of servicing or even replacing.

If you do not have isolating valves on either side of your pump, repairs or replacement will mean draining down the whole heating system. Having done that, take the opportunity to fit valves so that the pump can be more easily isolated for any future maintenance work.

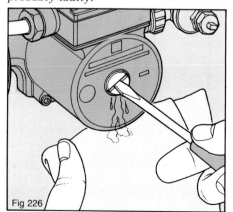

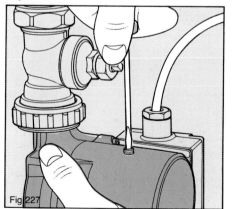

Fig 226   With many pumps, you can free a jammed impeller by inserting a screwdriver in a slot on the end of the pump casing, or by turning a knob in the same place until you feel the impeller move freely.

Fig 227   Bleed air from the pump by opening the bleed screw and listening for the sound of air escaping. Close the screw again when water appears from it.

Corrosion-Proofing and Descaling

Corrosion and scale are the two biggest enemies of any conventional wet central heating system. Corrosion occurs within the system by electrolytic means, caused by the presence of different metals (copper pipes and steel radiators), and the result is the gradual eating away of the steel to form iron oxides – the black sludge so familiar to any do-it-yourself plumber. The process causes leaks and blockages around the system, and is a major contributor to pump failure. Fortunately, it can be prevented by the addition of chemicals to the system.

Scale build-up within the system is also a problem, especially in hard water areas. Every conventional wet system takes in small amounts of fresh water from time to time, and this introduces a fresh supply of mineral salts which build up as scale deposits in the boiler and the system pipework. The result is noise and a gradual loss of boiler efficiency. Again, chemicals can cure the problem.

What to do

You need to descale the system thoroughly, then drain it down, flush it through with clean water and refill it again with fresh water to which a corrosion inhibitor has been added. Attach a hose to the system draincock, open the valve, draw off two or three buckets of water and close it again. Now add the descaling liquid to the feed-and-expansion tank as per the maker's instructions (see Check), and run the system to circulate it thoroughly.

Switch off the boiler, programmer and immersion heater, turn off the stoptap supplying the feed-and-expansion tank and open the draincock. As the system empties, open all the radiator air vents, working from the top downwards. Refill the system, pump it round, drain it down again and refill with fresh water (see Tip), closing air vents one by one as radiators fill up. Finally, add the corrosion inhibitor at the feed-and-expansion tank (see Check).

What you need:
- descaler
- corrosion proofer
- garden hose
- draincock key
- radiator key

CHECK
- that you use the correct chemical for descaling your system if it is more than 12 years old, because of the risk of leaks occurring as the scale is dissolved
- that you use the correct corrosion proofer; there are different types for cast iron or steel boilers and for copper tubular types, and another type specially formulated for use on systems with aluminium radiators. If you are unsure what you have, get expert advice.

TIP
When refilling the system for the last time, do it via the draincock rather than the feed-and-expansion tank to lessen the risk of airlocks. You must use a double-check valve on the hose supply to prevent back-siphonage.

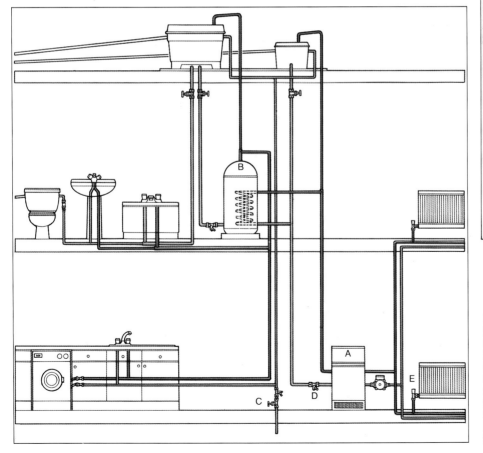

Fig 228 To drain down the heating system, switch off the boiler (A) and immersion heater (B), and turn off the stoptap supplying the feed-and-expansion tank (C). Open the system draincock (D) and radiator air vents (E). Refill the system from the top or bottom (see TIP).

Insulating Tanks and Pipes

Heat is energy and energy is expensive, so it pays to conserve as much of it as possible in the home. As far as your plumbing and heating system is concerned, this means using insulation to reduce heat losses from those parts of the system that convert and store hot water (the system pipework and the hot water cylinder) so that the maximum amount of heat reaches the points where it is needed (principally the radiators, and also the heating coil in the hot cylinder). Good insulation will not only save you money in terms of heating costs; it also means you can use less powerful (and so less expensive) heating appliances to deliver the heat you need.

Insulation has another important purpose; protecting the system from freezing in cold weather. Parts of the system containing cold water in areas such as lofts and underfloor voids are at risk here, and good insulation of supply and circulation pipes and cold water tanks is essential.

What to do

The two most important parts of your system to attend to are the hot cylinder and any pipes and tanks in the loft, especially if this is well insulated.

Fit a proprietary jacket to your hot water cylinder, even if this already has pre-shrunk foam insulation. Do not cover the immersion heater thermostat or a cylinder stat if one is fitted.

In the loft, either fit a proprietary jacket (available for most sizes of round plastic cistern) or use loft insulation materials. Wrap glass-fibre blanket round the tanks, or make a box of rigid polystyrene to fit round them. Insulate all pipework in the loft space using pre-formed lengths of insulation to suit the pipe sizes, and also all supply pipes run beneath timber ground floors. Wherever possible, insulate all central heating pipework wherever it runs, to reduce heat losses from the system.

What you need:
- cylinder jacket
- jackets for cold water storage tank and feed-and-expansion tanks in the loft *or*
- insulation materials for wrapping tanks
- pipe insulation to match pipe sizes
- tape or string

CHECK
- that there is no insulation beneath tanks in the loft, so that the small amount of heat rising from the house below can reach them and help prevent a freeze-up
- that insulation on cold water storage tanks or feed-and-expansion tanks does not interfere with the vent pipes discharging over them
- that any pipes running next to outside walls or close to air bricks are insulated

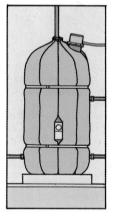
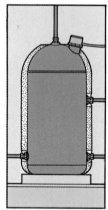
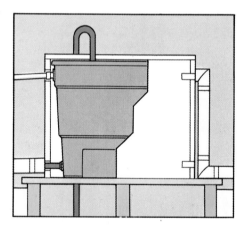

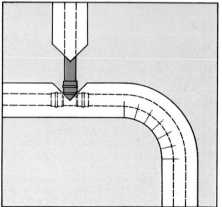
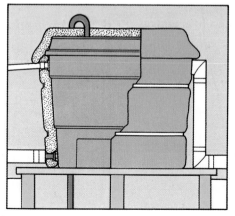

Fig 229 Insulate cold water storage tanks and feed-and-expansion tanks in the loft with rigid or flexible insulation materials, or fit a proprietary jacket to round plastic tanks. Use proprietary jackets on hot cylinders, even if they have pre-shrunk insulation already. Use foam insulation on pipework, taping bends and shaping tee connections neatly.

EXPENSIVE JOBS

The jobs described earlier in this book are all well within the abilities of the average do-it-yourselfer working on a new or reasonably modern plumbing system. However, there are still many homes in this country plumbed with materials such as iron or lead which are difficult for the amateur to work with, or fitted with old-fashioned and out-of-date equipment which is often very awkward to repair or maintain satisfactorily. It is generally best to leave such tasks to an expert rather than to attempt to tackle them yourself.

The same point applies to large-scale projects such as replacing the existing plumbing system, changing central heating boilers and altering or extending existing heating systems. Some do-it-yourselfers do tackle them, but most prefer not to.

Old Plumbing Systems

You'll know if you have an old plumbing system. It will clank and groan its way through the day, plagued by dripping taps, WCs that don't flush properly and even the occasional leak to keep you on your toes. It may be more than a source of irritation; it may also be bad for you. There is mounting evidence of the harmfulness of lead in drinking water, and many old systems still contain large amounts of lead piping. Furthermore, inadequately protected storage tanks in poor condition may allow contamination of stored water to occur, and drips and leaks will cost you money as hot water runs needlessly to waste. Updating such a system need not cost a fortune, and will radically improve your quality of life.

Old Heating Systems

Similar criticisms apply to old heating systems (and some 30 per cent of our housing stock still has no central heating, relying instead on solid fuel fires and an assortment of individual heaters for warmth and hot water). Modern boilers are far more efficient than those made even as recently as ten years ago, and changes in the building regulations now permit the use of unvented heating sys-

tems which greatly simplify the plumbing requirements because they do not require tanks in the loft (although they must be professionally installed). Modern radiators are more efficient too, and increasingly sophisticated system controls now allow you to regulate how your system operates much more effectively than ever before. In addition, better standards of insulation in homes allow smaller boilers and heat emitters to be specified, reducing capital expenditure as well as running costs.

Old Wastes and Drains

The last part of your plumbing system that may need attention is how it gets rid of waste water. Many older homes still have two-pipe drainage systems, with waste water running into hoppers and gullies which frequently become blocked and can be a source of smells. Waste pipes may be of lead or other metal, with traps that are difficult to clean and pipes that are inaccessible if blocked. Here the fitting of modern waste pipes, plus alterations to the way waste water reaches the drains, may be of major long-term benefit.

Fig 230 (*above*) Major projects such as changing boilers, installing central heating and replacing plumbing supply pipework and waste systems are best left to professional installers unless you are a very experienced amateur plumber.

Planning a Major Project

One of the main reasons why many people decide to leave major plumbing projects to an expert is the time factor. It is difficult to carry out the work without major disruption to the house and its services, and experts can both work faster than even an enthusiastic amateur, and use their skill and knowledge to keep such disruption to a minimum. However, choosing this route does not mean that you cannot take an active part in deciding what you want your project to achieve. Here are some of the points you should be thinking about.

Reorganizing Things

For projects such as plumbing in a new bathroom suite or installing a new central heating system, don't just swap like for like. In the bathroom, for example, it may be possible to arrange the various pieces of equipment in a different way, perhaps allowing for the use of a corner bath instead of a conventional one, or the installation of a shower instead of a bath to free extra floor space for a bidet. Houses are permitted by the new building regulations to have a bath or a shower, and there is no doubt that the latter will save money and time as well as being more hygienic; it all depends on whether you are prepared to forego that long soak at the end of a hard day.

You could even consider changing the location of the bathroom completely – if this does not involve major problems in getting rid of waste water. The bathroom in many older houses is huge, and it may be practical to use it as a bedroom instead and site a new bathroom elsewhere – perhaps partitioned off an existing room.

As far as central heating is concerned, modern wall-mounted gas boilers can be placed almost anywhere in the house or in an adjoining garage, perhaps freeing space currently occupied by a larger floor-standing boiler or allowing a fireplace to be opened up again. Modern radiators are smaller, neater and less obtrusive than their older counterparts, and can be sited away from windows if desired. Pipework can be run round rooms behind hollow skirting mouldings, so minimizing disruption and making future maintenance or repairs far simpler to carry out than if they are buried under floors.

Using the Latest Fittings

Don't be afraid to ask what's new in the plumbing business when talking things over with your plumber. At the most basic level, using one of the recently-approved plastic pipework systems could speed up the installation work considerably, and leave you with a system that will be much easier to modify and extend in the future.

Taps with ceramic discs instead of tap washers will mean an end to rewashering (although they do not last for ever without needing some attention), and are simpler to operate, especially for children and the elderly or handicapped. Modern ball valves operate almost silently, and are far easier to adjust and maintain than their old-fashioned forerunners.

Bathroom and kitchen equipment is now getting far more attention than ever before, with innovative designs and unusual features becoming more common. However, there are one or two pitfalls to watch out for. For example, many modern sinks are too small to accept things like roasting trays and grill pans, so decide first if this is important to you. If you live in a hard water area, don't choose bathroom sanitaryware in deep colours, or you will be forever battling against visible scum and scale marks. And think twice about having a Jacuzzi unless you have a water softener; keeping all those nozzles clean and working could become a nightmare if you live in an area with hard water.

On the heating front, ask whether a combination boiler or a condensing boiler is worth considering. The former provides hot water as well as heating the radiators, so does away with the need for a hot water cylinder, although it can generally supply only one hot tap at a time. The latter is much more efficient than a conventional boiler, especially when working at low output in hot weather, although it costs about 30 per cent more to buy than a conventional boiler. They can be fitted only to fully-pumped gas-fired systems.

Lastly, ask about controls. You can have separate heating zones, independent control of heating and hot water and separate temperature control of each room, and even sensors that take account of outside temperature changes and adjust the heating accordingly.

CHECK
- that your installer is qualified and competent. Personal recommendation is best, but otherwise approach installers who are members of the Institute of Plumbing, the Confederation for the Registration of Gas Installers (CORGI), the National Association of Plumbing, Heating and Mechanical Service Contractors, the Heating and Ventilating Contractors' Association (HVCA), the Solid Fuel Advisory Service (SFAS) or the Scotland and Northern Ireland Plumbing Employers' Federation. Most operate codes of practice or warranty schemes for your protection. (See page 94 for addresses.)

TIP
When you are having extensive plumbing or heating work done, get your installer to fit appropriate isolating valves and draincocks at every sensible point on the system, to make any future maintenance or alteration work easy to carry out.

FACTS AND FIGURES

This section is intended as a handy reference guide to the range of plumbing pipes, fittings and materials you will need for various plumbing and heating jobs around the house. It will help you to see at a glance what is available, and what to use where, so you can plan your requirements in detail and draw up shopping lists for individual projects.

It also contains diagrams showing the most widely used hot and cold water supply, and waste water and central heating system layouts, so you can refer quickly to them while you work instead of having to leaf backwards and forwards between subjects that are dealt with elsewhere in the book.

Lastly, on page 94 there is a detailed glossary of all the terms used in the book, plus a list of useful addresses of relevant trade organizations.

Pipe Types and Sizes

Below is a check-list to help you choose pipework of the right type and specification for each job.

Copper pipe Use copper pipework for all general-purpose plumbing and heating work if your home already has it and you feel confident about cutting, bending and joining it. Make sure it is marked as conforming to British Standard BS 2871 Part 1 Table X. It is generally available in 3 and 6m lengths. The sizes you need are:

15mm – for all mains-pressure and most low-pressure plumbing and heating runs.
22mm – for feed pipes to hot cylinders and bathrooms (including bath taps), for most boiler primary circuit pipework and for parts of the heating system.
28mm – for some main boiler pipework if required by the system design.

Plastic pipe Three types of plastic pipe – flexible polyethylene (Pipex), flexible polybutylene and rigid cPVC (Hunter Genova) – are approved by the Water Research Centre for use on domestic plumbing supply and central heating pipework. All are available in 15 and 22mm sizes, in cut lengths of 2 or 3m and in longer coils in the case of polybutylene

and polyethylene pipe, which are both flexible enough to be bent round corners without the need for elbows.

Plastic waste pipes are made from ABS, polypropylene and uPVC (the last two are also used for overflow pipes). Standard sizes are 22mm (overflow only), 32mm, 40mm and 50mm; standard lengths are 2, 3 or 4m. Note that individual brands are often not compatible with different makes.

Plastic soil pipes are made from uPVC. Standard sizes are 82mm (seldom used in the home) and 110mm; standard lengths are 2, 3 and 4m.

Rain-water systems Plastic gutters and downpipes are made as complete systems, and components are not generally interchangeable between brands. Common nominal sizes are 75mm half-round gutter with 50mm round pipe; 100mm square-section gutter with 60mm square pipe; 110mm half-round gutter with 65mm round pipe; 150mm half-round gutter with 100mm round pipe. Standard lengths for both gutters and downpipes are 2, 3 and 4m.

Fig 231 (*above*) Whatever project you are working on, draw some simple plans first so you can work out in detail what materials you need.

Plumbing Fittings

There is a huge range of plumbing fittings available, and it can often be highly confusing trying to choose the correct one for a particular job. Below is a summary of the main types you are likely to need.

Which Material?

For supply pipework, your choice is between:

Copper capillary fittings – for making soldered joints in copper pipe.

Brass compression fittings – for making joints in copper, polythylene or polybutylene pipes using spanners.

Push-fit fittings – also for making joints in copper, polyethylene or polybutylene pipes, connected by hand.

Solvent-weld fittings – for joining rigid cPVC pipes.

Which Fitting?

You need fittings for connecting pipe to pipe, and pipe to other components of your plumbing system.

Supply pipe connectors The basic connector is the *straight coupling*, used to link lengths of pipe in a straight line. This is available in all materials, in sizes to link pipes of the same size or, in the case of reducing couplings, pipes of different sizes.

For turning sharp corners you need *90° elbows*, while for connecting branch pipes you need *tees*. These are available with all ends equal, or with one or two of the three outlets reduced to allow pipes of different sizes to be connected in different configurations. Elbows are also available with draincocks incorporated.

Connectors to appliances For connecting pipes to the threaded tails of taps and ball valves, you need a fitting called a *tap connector*. This incorporates a washer and coupling nut, and is available in ½ and ¾in sizes in all materials in straight and 90° elbow versions. Reducing connectors are available for linking 15mm mains-pressure pipework to ¾in bath taps, and for connecting 10mm mono-bloc tap tails to 15mm supply pipes.

For connecting pipes to appliances such as boilers, hot cylinders and storage tanks you need *fittings with one end threaded*. Fittings with an external thread are described as 'male iron' because they have a

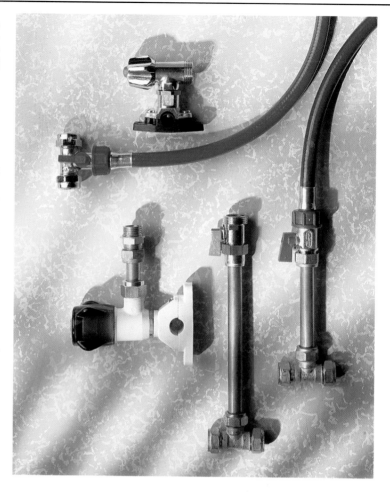

British Standard Pipe (BSP) thread pattern that matches that used on iron pipework. Similarly, fittings with an internal thread are described as 'female iron'. Both types come in all common sizes with compression, capillary, push-fit or solvent-weld connections to the incoming pipework, and may be straight couplings or 90° elbows.

System controls For isolating the rising main and any other mains-pressure pipework, you need *stoptaps*. Most domestic supply pipework is run in 15mm pipe, so you will need just the one size. Note that stoptaps must be fitted the right way round so the flow direction matches the arrow on the tap body. You may also need mini-stoptaps to allow the supply hoses of washing machines and dishwashers to be connected to their pipework.

For isolating other parts of the system you need *gate valves* – the 22mm size on

Fig 232 (*above*) All supply pipe fittings you use must comply with the requirements of the Water Supply By-laws. They must bear the British Standards Institute Kitemark, or the Water Industry Fittings Testing Scheme mark – signified by a water droplet with an L superimposed on it.

Plumbing Fittings

feeds from the cold water storage tank, and smaller 15mm ones for supply pipes to individual appliances.

You will need *draincocks* at various points on the plumbing and heating systems to allow them to be drained down for repair and maintenance work or alterations. Both 15mm and 22mm sizes are readily available.

The new water supply by-laws require the use of a *double check valve* at certain points on the plumbing system to prevent any risk of back-siphonage of stored or used water should mains pressure drop for any reason.

For controlling conventional heating radiators you need *lock-shield* and *hand-wheel* (or *thermostatic*) *radiator valves*, which are available in a standard size to connect the 15mm supply pipework to radiator couplings.

Waste pipe fittings For waste pipes and overflows, your choice is between *push-fit* and *solvent-weld connectors*. The former incorporate a sealing ring within the fitting, and allow pipe runs to be dismantled easily for maintenance. They can be used with any type of plastic piping, but care must be taken to ensure that pipe and fittings are compatible; external pipe diameters vary slightly from brand to brand, even though internal pipe diameters are standardized.

Solvent-weld fittings are less obtrusive than push-fit ones but joints are permanent once made and pipe runs must incorporate expansion joints at intervals. They cannot be used with polypropylene pipes. Again, pipe and fittings must be compatible.

Straight couplings, 90° and 135° elbows, swept elbows, swept tees, reducing couplings and of course traps are available in both push-fit and solvent-weld ranges, to match pipework in the four standard diameters – 22mm for overflow pipes, 32mm for basins, 40mm for baths, showers, sinks and washing machines or dishwashers, and 50mm for pipe runs over 2m (6ft) long.

Soil pipe fittings Almost all soil pipe fittings are for making solvent-welded joints, for obvious reasons. There is a huge range of fittings available to enable soil stacks to be run in almost any direction and to get round obstacles in their path. The range includes straight couplings, expansion fittings (push-fit, not solvent-

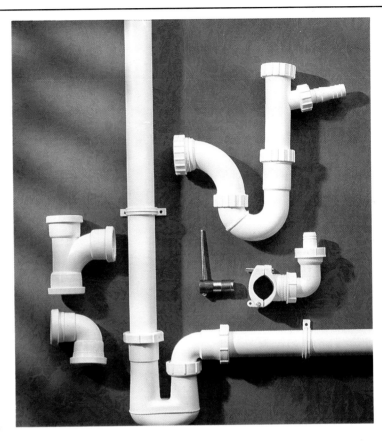

weld), 45°, 90° and 168° elbows, variable bends, branch and boss connectors and new multi-connection manifolds; the last allows several pipes to be connected at the same point without the risks of back-siphonage present with conventional connections. There is a standard nominal pipe size of 110mm for soil pipe, but as with waste systems it is essential to use pipes and fittings that are compatible with each other, which generally means sticking to one brand wherever possible.

Push-fit connections are almost universal nowadays for connecting WC pan outlets to soil pipe runs, and several ranges of proprietary connectors are available to cope with all possible permutations of pan and pipe size, and position.

Fig 233 (*above*) Waste and soil pipe fittings must comply with the building regulation requirements. Different brands may not be compatible with each other.

Hot and Cold Water Supply

On a typical installation, the rising main supplies the kitchen cold tap (and perhaps also branches to a washing machine, a dishwasher and an outside tap), and then rises through the house to fill the cold water storage tank and the separate small feed-and-expansion or header tank. Both these tanks have overflow pipes leading to the eaves (Fig 235).

From the cold water storage tank, one feed pipe runs to the hot cylinder and another supplies all the other cold taps and the WC cistern(s). Each WC cistern also has an overflow pipe.

The hot cylinder is heated by gravity circulation to and from the boiler, and this circuit is topped up where necessary by water from the header tank, which also accommodates expansion in the system when it is hot. Hot water is drawn off at the top of the cylinder and taken to the hot taps. A safety or vent pipe is also connected at this point and discharges over the cold tank. A similar vent pipe runs from the boiler circuit to discharge over the header tank (Fig 234).

Some homes may have direct cold water supplies to taps and WC cisterns throughout the house, fed by branches off the rising main.

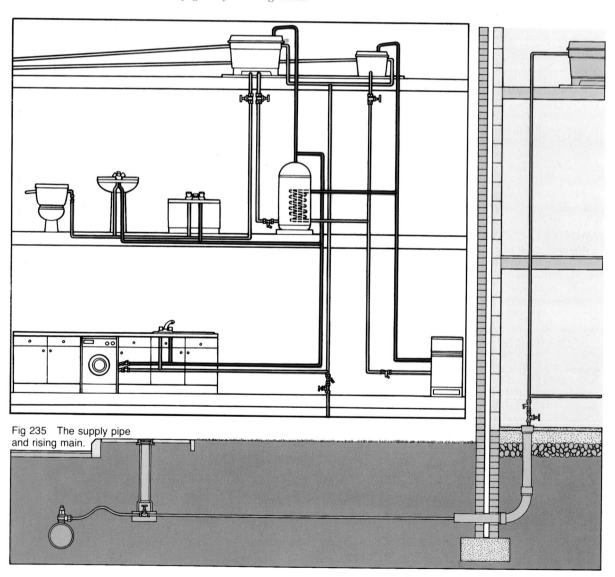

Fig 235 The supply pipe and rising main.

Waste and Soil Pipe Systems

Modern homes have a single-stack drainage system, with the WC and most other appliances discharging their waste water directly into it. The main connection is the WC branch, and no other waste pipe may be connected to the stack within 200mm (8in) of this point. The stack is generally vented above roof level, but stub stacks may be terminated within the house so long as they are fitted with a pressure relief valve. Rodding eyes are generally fitted at various points to allow blockages to be cleared easily. At the foot of the stack a slow bend connects the stack to the underground drains. Inspection cham-

bers on modern systems are generally made of plastic (Fig 236).

Other appliances remote from the main stack discharge their waste via back-inlet gullies, with the pipes taken below the level of the gully grid. Rain-water downpipes discharge into separate gullies leading to surface-water drains or soak-aways.

Older homes have a two-pipe drainage system. Waste from the WC goes into the soil pipe, while waste from all other appliances discharges into wall-mounted hoppers upstairs and ground-level gullies otherwise. Rain water downpipes often discharge into the same gullies, or may run to separate soak-aways (Fig 237).

TRAPS
Traps are fitted to all water-using appliances and contain a water seal to prevent drain smells from entering the house. They are also built into underground gullies, and may be found in the last inspection chamber on old drain runs.

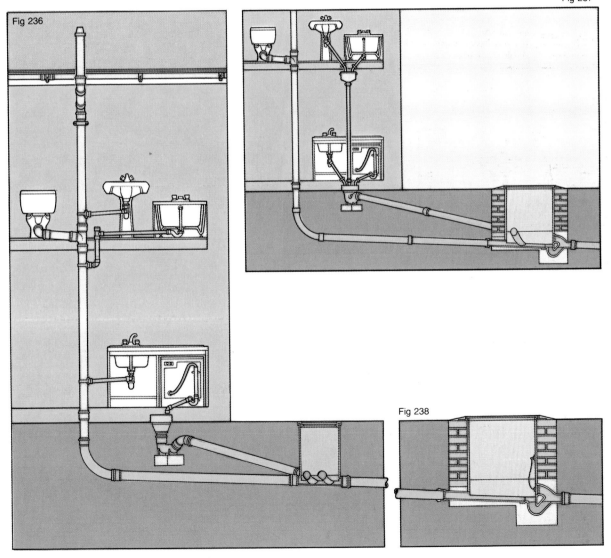

Fig 236

Fig 237

Fig 238

Heating Systems

Many homes have a gravity-circulation hot water system and a fully pumped central heating system. The system controls can be set to provide hot water only, or both water and space heating, but not space heating without water heating. There may be a towel rail in the bathroom which is connected to the flow and return pipework between the boiler and the hot cylinder (Fig 239).

The heating circuit usually has the pump on the return side of the pipework, and its operation is controlled by the room thermostat and the system programmer. Each radiator is connected to the flow and return pipework and may be controlled by a manual or a thermostatic radiator valve.

Some homes have a more sophisticated arrangement with both heating and hot water circuits being fully pumped. This results in faster heating of the hot cylinder than is achieved with the conventional method of gravity circulation, and is useful in homes where there are short periods of high water demand. It also allows the boiler and hot cylinder to be mounted far apart with little loss of efficiency. A motorized valve diverts water to the hot water or heating circuits under the control of the system programmer, and zone valves can allow the creation of separate heating levels for different parts of the house.

SAFETY FEATURES

The main safety features of both these systems are the vent pipes discharging over the cold water storage tank and the header tank. In addition, a pressure relief valve may be fitted to the primary circuit pipework. On no account must a gate valve be fitted to the primary circuit

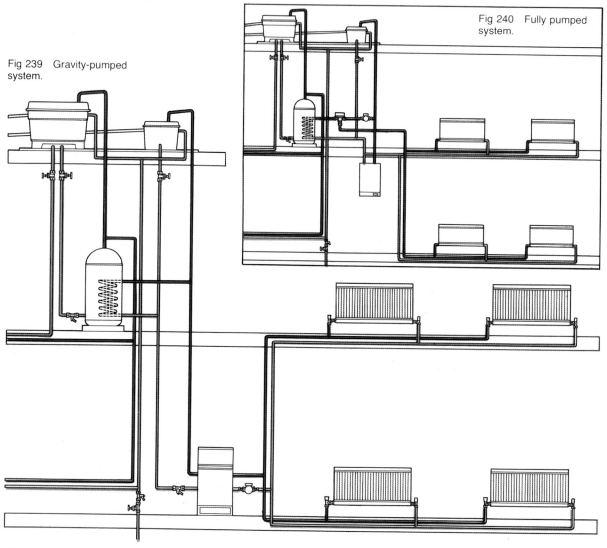

Fig 239   Gravity-pumped system.

Fig 240   Fully pumped system.

Heating Systems

Mains-fed Water Heating

Where there is no room for hot water storage – in flats, for example – a mains-fed system may be installed instead. Here mains-pressure water is supplied to cold taps and WC cisterns, and also to a multi-point water heater which supplied the hot taps. A similar arrangement can be used to supply central heating as well, with a combination boiler taking the place of the multi-point heater.

Unvented Systems

Recent changes in the building regulations now allow the installation (by professional fitters only, and under closely controlled conditions) of unvented water heating systems which avoid the need for storage and header tanks and provide both hot and cold water at mains pressure. The system incorporates a number of built-in safety features.

Thermal Store Cylinder

This is an alternative to unvented systems, giving the benefit of mains-pressure hot water. Incoming cold water is heated rapidly by passing it over a very efficient heat exchanger inside the cylinder, whose contents remain captive within the boiler and heating circuit.

Fig 241 Mains-fed hot water supply via a multi-point heater or combination boiler.

Fig 242 Unvented high-pressure water heating system with no storage or header tanks.

Fig 243 Thermal store cylinder, with incoming cold water heated by passing it through a coil in the cylinder.

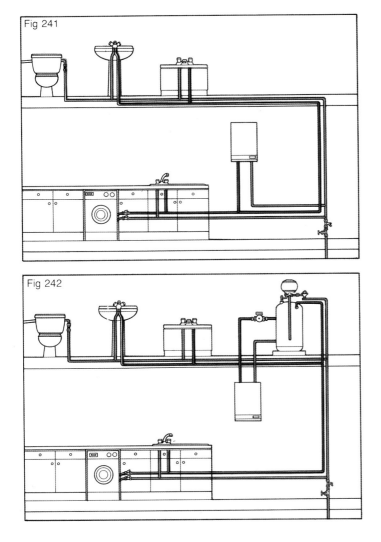

Fig 241

Fig 242

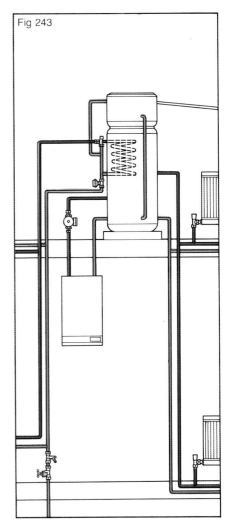

Fig 243

Glossary

Air vents   Special valves fitted to radiators and at other points in the system to allow trapped air to be removed.

Ball valves   Float-operated valves fitted to storage and WC cisterns to allow them to refill automatically.

Bending springs   Flexible springs used to support pipe walls and prevent kinking during pipe-bending operations.

Capillary fittings   Copper pipe fittings used to make soldered joints.

Cisterns   Storage tanks for cold water. Small cisterns also supply WCs.

Compression fittings   Brass pipe fittings used to make mechanical joints on pipework. They are tightened with spanners.

Crowsfoot spanners   Open spanners with the jaws at right angles to the shaft. They are used to connect and disconnect pipework to taps in confined spaces.

Cylinders   Storage vessels for hot water, supplied in most systems from the loft storage cistern. Most contain a heat exchanger linked to a boiler, or an electric immersion heater.

Draincocks   Small taps fitted at low points on plumbing systems to allow pipes to be drained for maintenance or repairs.

Feed-and-expansion tanks   Fitted to hot water and heating systems to replace any water losses and cope with expansion within the system.

Gate valves   Valves fitted to low-pressure pipe runs to allow them to be isolated for repair or maintenance.

Gullies   In-ground collection points for water from indoor appliances and downpipes. They contain a trap to prevent drain smells from entering the house.

Hoppers   Above-ground collection points for water from indoor appliances installed at first-floor level. They are fixed to the outside wall of the house and are linked to a gully by a downpipe.

Immersion heaters   Electric water heaters installed in hot cylinders to provide a supply of hot water, either on their own or in tandem with a boiler.

Inspection chambers   Better known as manholes, they are fitted to drain runs where new drains join, or at changes of direction, to allow access to the drains.

Mixer taps   They have twin inlets for hot and cold water, and one outlet nozzle which may be fixed or movable. In kitchen mixers the flows are kept separate until the water leaves the nozzle.

Mono-bloc taps   Mixer taps designed to fit a single mounting hole in a sink, basin or bidet. The slim pipe tails are linked to the supply pipework with reducing couplers.

Motorized valves   Used on heating systems to divert the flow of hot water to different parts of the system.

Pillar taps   These have a vertical water inlet, and are fitted to baths, basin, sinks and the like.

Pipe clips   These are made of metal or plastic, and are used to secure pipework.

Rising main   A term used to describe the incoming mains-pressure supply pipe, which enters the house and rises to the storage cistern in the loft.

Single-stack drainage systems   These take waste water from indoor appliances and WCs direct to the drains via a single stack, which can be installed inside the house.

Solvent-weld fitting   Used with a special solvent to make joints in some types of plastic pipework.

Stoptaps   Fitted to mains-pressure supply pipes to regulate the flow rate and to allow the supply to be shut off for maintenance or repairs.

Supataps   Taps with a washer and jumper which fall away from the tap seating as the tap is opened. They can be rewashered without turning off the supply.

Thermostats   Electrical devices that monitor temperatures on heating systems.

Traps   Fitted to all water-using appliances to keep drain smells out.

Two-pipe waste systems   These have separate pipes for waste water from appliances and soil water from WCs. They are now obsolete.

Vent pipes   These are fitted on vented heating and hot water systems as a safety valve in the event of the system overheating.

Washers   Those made of rubber are fitted inside most taps to make a watertight seal on the tap seating when the tap is closed. Many modern taps now contain ceramic discs instead.

Yorkshire fittings   Capillary fittings containing an integral ring of solder. They are simply heated with a blowtorch or special soldering iron to melt the solder and make the joint.

USEFUL ADDRESSES

Confederation for the Registration of Gas Installers (CORGI),
St Martins House,
140 Tottenham Court Road,
London W1P 9LN.

Heating and Ventilating Contractors' Association,
34 Palace Court,
London W2 4JG.

Institute of Plumbing,
64 Station Lane,
Hornchurch,
Essex RM12 6NB.

National Association of Plumbing, Heating and Mechanical Service Contractors,
6 Gate Street,
London WC2A 3HP.

Scottish and Northern Ireland Plumbing Employers' Federation,
2 Walker Street,
Edinburgh EH3 7LB.

Index

Index